His *and* Hers

His

Gender, Consumption, and Technology

Roger Horowitz and Arwen Mohun, Editors

and

Hers

University Press of Virginia
Charlottesville and London

Acknowledgments for previously published material appear on pages 182 and 222.

The University Press of Virginia

Printed in the United States of America

First published 1998

The paper used in this publication meets the minimum requirements of the American National Standard for Information Sciences–Permanence of Paper for Printed Library Materials, ANSI Z39.48-1984.

Library of Congress Cataloging-in-Publication Data

His and hers : gender, consumption, and technology / Roger Horowitz and Arwen Mohun, editors.
p. cm.
Includes bibliographical references and index.
ISBN 0-8139-1803-0 (cloth : alk. paper). —
ISBN 0-8139-1802-2 (pbk. : alk. paper)
1. Consumption (Economics)—Social aspects—United States—History. 2. Consumer behavior—United States—History. 3. Women consumers—United States—History. 4. Technology—Social aspects—United States—History. I. Horowitz, Roger. II. Mohun, Arwen, 1961– .
HC110.C6H552 1998
306.3'4'0973—dc21 98-10278
CIP

Contents

Illustrations

His *and* Hers

Introduction

Roger Horowitz and Arwen Mohun

His and *hers, consumption,* and *technology*—all words suggestive of a complex historical phenomenon. The chapters in this collection are among the first to explore the history of consumption by synthesizing discrete historical literatures on consumer culture, gender, and the history of technology. Collectively they demonstrate the value of analyzing consumption as both a material and a cultural process. Production and consumption are equally inseparable. Tools from both the history of technology and gender studies illuminate how these categories intersect. Although broad social and technological trends influence the outcome of these stories, the authors emphasize the agency of particular groups, including consumers, workers, manufacturers, and the "mediators" who communicated between producers and consumers.[1]

Technology played a crucial role in the development of a modern consumer culture. These chapters define technology broadly—as discrete commodities, such as radios and flatirons, but also as technological systems, such as hotels, experienced by consumers as environments or services. Incorporating technology restores the material dimension of consumption and reveals the limits of treating technology connected to consumption as an unopened "black box" not requiring further investigation.

Producers and consumers in these chapters are understood to be makers and users of technology. Consequently, it becomes clear that consumers make choices based not just on what technology *means* but also on how it *functions.* Similarly, producers must reconcile what they think consumers want with the economic and technological constraints of manufacturing objects or shaping physical spaces. The insights gained by examining the linkages between production and consumption illustrate the interrelationship between them and the problems of analyzing each separately.

Gender also matters. But how it matters, as the following chapters show, can vary enormously. In some cases gender shaped the physical structure of sites of consumption, such as the nineteenth-century luxury hotel and the mid-twentieth-century shopping center. In other instances gender relations influenced the development of entirely new products, such as inexpensive chocolate candy and factory-made glassware. Gender also informed marketing strategies and assumptions (often fallacious) of male designers and the stubborn preferences of female consumers. These women, as historians of technology have shown, are technological actors rather than passive objects of marketing efforts. These chapters use gender analysis to build on this observation. Not only do they show that consumers are important agents in the development of technologies but they explain how stereotypes of women as passive consumers and men as producers have been historically constructed. The authors explore how this process could reinforce cultural notions about appropriate roles of behavior for men and women while providing opportunities for them to refashion their identities.[2]

Despite the wide range of temporal and topical foci, several underlying themes emerge from these chapters. Collectively they portray consumption—and the making of commodities for consumption—as resulting from complex negotiations between consumers and producers. They interrogate what Ruth Schwartz Cowan has called the "consumption junction," exploring how makers of goods worked hard to ascertain consumer preferences and to find customers for their products. Rather than seeing consumption as a story of either resistance or dominance, the authors conceive of consumption as an exacting process in which firms and consumers struggled to communicate—and sometimes failed to do so successfully. The chapters seek to unravel how producers of goods and services learned approaches and developed techniques to convince individuals to consume, especially when consumers insisted these products fit into their lives and satisfy their desires. As they do so, the authors often differ about the relative power of producers and consumers.[3]

In their exploration of the culture of consumption these chapters make a special contribution by identifying agents of cultural mediation. These "translators" were individuals or institutions who facilitated communication between consumers and producers. They range from home economists who worked for electric companies to buyers in department stores who advised china and glass manufacturers. Without these translators firms could and did make products that consumers would not buy.

These chapters also tell us how firms and their translators tried to understand, appropriate, and alter popular conceptions of leisure, pleasure, and utility. Courtship and the family occupy a special place in these efforts, befitting the gendered character of consumption. Hotel operators and chocolate manufacturers sought to develop entirely new markets by tapping into basic aspirations for luxury and enjoyment and developing new forms of consumption habits out of these desires. Electric energy producers and shopping mall builders tried to alter established practices of household work and shopping in order to expand their markets.

These initiatives, however, were fraught with errors and false starts. Manufacturers of glassware and electrical appliances struggled to develop products that fit with existing consumption patterns and consumer aspirations. Time and again firms had to employ translators to interpret the mysterious codes of consumption and articulate how firms could sell their products.

The chapters begin with a broad historiographic and thematic overview by Steven Lubar and then proceed chronologically through a series of case studies. Lubar's chapter establishes the broad contours of how scholars have linked gender to the production and consumption of goods and services in the United States since the mid–nineteenth century. He stresses that gender applies to both men and women, whereas traditional studies of technology have focused on the manly, production side of the circulation of commodities. Lubar also contends that isolating production from consumption has resulted in an incomplete, hence inaccurate picture of the development of the market and modern capitalist society. He closes by identifying promising lines of inquiry that address this imbalance, including work by some of the authors in this collection.

Molly Berger's chapter traces the dynamic interaction between gender, consumption, and technology to antebellum America and the rise of the urban luxury hotel. These hotels were quintessential sites of consumption and production because providing guest services rested on complex internal technological systems and a substantial staff. As such, these institutions fused—not without tension—the separate male and female spheres of nineteenth-century elite society. While forms of "technological luxury" in hotels reinforced male notions of progress, these systems were located in a luxurious domestic environment saturated with femininity. Consequently, space within the hotel was allocated by sex; men received the bulk of the public space, with their areas fashioned in a manner to accommodate masculine habits. However, the opportunity for women to enter a regulated, commercial public space was still an unusual opportunity, and many younger women seized it as an opportunity to meet marriageable young men.

Following Berger's chapter, several chapters continue this story into the late nineteenth and early twentieth centuries. Gail Cooper relates how chocolate changed from a ubiquitous symbol of romance and courtship in the late nineteenth century to an everyday indulgence after World War I. This transformation entailed changes in production methods and consumption habits and inspired female reformers to use women's power as consumers to influence employee relations in the candy industry. The impact of gender and the close relationship between production and consumption of candy validate Cooper's contention that using "the chocolate bonbon, instead of the automobile," to understand America's consumer society in the first third of the twentieth century has immense benefits.

James Williams shows how gender informed California power companies' efforts to make better use of their ample electric generating facilities by persuading consumers to acquire electrical devices. To evaluate gender's effect on the dissemination of newly developed household technologies, Williams contrasts successful efforts to sell housewives simple appliances like electric irons with the failed attempt to market more complex electric ranges. He attributes these results to misconceptions rife in male marketing departments concerning women's actual work in the home, especially men's failures to appreciate that women preferred to incorporate new technology into existing patterns of daily activities. The limited successes of these firms resulted from their employment of female "translators" to sell electrical devices and to create sales environments and displays welcoming to female customers.

In a chapter that focuses on the 1920s and 1930s Louis Carlat shows how manufacturers changed the character of radios to appeal to female consumers. Initially these devices' complexity attracted male tinkerers, men who devoted long hours to the intricate task of tuning components to receive radio waves. Altering radio's gendered appeal necessitated transforming the device from a "male toy" relegated to the workshop to a "feminine object" best located in the domestic space of the parlor. In so doing manufacturers changed radio's function from conquering distance to protecting and indeed invigorating the private space of the home.

Regina Blaszczyk's chapter explores the centuries-old "Cinderella paradigm," which established glassware and other fine dining objects as central to female domesticity and refinement. Beginning with the interwar years and moving into 1950s America, she shows how this paradigm provided a set of ideas and desires that modern glass manufacturers used to sell their products. Her story focuses on the efforts of the Fostoria Glass Company to market glassware, especially to brides. In this study of manufacturers' perceptions of consumers Blaszczyk demonstrates how in the "gendered marketplace" of glass "producers yielded to their audiences" while seeking to translate the "dreams" of consumers "into reality as beautiful, fragile glass objects."

Joy Parr looks into the post–World War II expansion of consumption in her parable "Shopping for a Good Stove." She investigates the contradictions created when men designed stoves for women and devised marketing strategies to sell them to female shoppers. In this tale men's gendered assumptions interfered with and indeed crippled effective design, marketing, and sales strategies. Parr shows how male stove designers and salesmen created goods and sales campaigns that clashed with the actual process of preparing food and did not engage the preferences of the women to whom they were selling.

Finally, Lizabeth Cohen addresses the impact of the shopping center, the new forum for modern consumption that initially emerged in the 1950s and is so commonplace today. She shows how the privatization of formerly public commercial space—the shopping district—dramatically changed the experience of consumption in suburban New Jersey. With the suburban middle-class and male-breadwinner family as their model, shopping-center designers planned their new sites to appeal to a feminine and class-segmented audience. This was a great irony, for women's enhanced "claim on the suburban landscape" remained limited to the realm of consumption despite their increasing role on the sales staffs of shopping-mall retail stores.

Cohen's conclusion takes us back to one of Lubar's main points in the opening chapter: the gendered construction of the production-consumption dyad in modern capitalist society. As these chapters show, patterns of consumption have changed a great deal since the days of antebellum luxury hotels, but the importance of gender and the role of women in consumption have persisted.

This book can be traced to an April 1994 conference, "His and Hers: Gender, Technology, and Markets," sponsored by the Center for the History of Business, Technology, and Society at the Hagley Museum and Library in Wilmington, Delaware. The conference and this book would not be possible without the Hagley's support and the efforts of its staff, especially the Center's director, Philip Scranton, and coordinator, Carol Ressler Lockman. Former Center associate director Julie Johnson deserves special thanks for first proposing the "His and Hers" conference.

Notes

1. T. Jackson Lears, *The Culture of Consumption: Critical Essays in American History, 1880–1980* (New York: Pantheon, 1983). An otherwise excellent book that typically neglects technology is Victoria de Grazia, ed., *The Sex of Things: Gender and Consumption in Historical Perspective* (Berkeley: Univ. of California Press, 1996).

2. Ruth Schwartz Cowan, *More Work for Mother: The Ironies of Household Technology from the Open Hearth to the Microwave* (New York: Basic Books, 1983).

3. Ruth Schwartz Cowan, "The Consumption Junction: A Proposal for Research Strategies in the Sociology of Technology," in *The Social Construction of Technological Systems: New Directions in the Sociology and History of Technology*, ed. Wiebe E. Bijker, Thomas P. Hughes, and Trevor J. Pinch (Cambridge: MIT Press, 1987), 261–80.

One

Men/Women/ Production/ Consumption

Steven Lubar

It is easy to assume that business and technology are beyond gender, driven by markets and mechanics and not by culture. This chapter begins with the opposite assumption, that both production and consumption are gendered. It considers some of the ways in which cultural ideals of what is appropriate for men and women have influenced design and manufacture, buying and selling, the ways we make and distribute and use things. How, it asks, have business and technology reflected, constructed, maintained, and reinforced ideas about gender?

In this chapter I outline some of the ways in which considering gender allows us a more complete understanding of the workings of markets and machines. Judy McGaw, who has written some of the most insightful analyses of gender in the history of technology, suggested this analysis in an essay outlining "a feminist perspective on technology's history." "We will do a better job of examining women, men, and technology historically," she wrote, "if we heighten our consciousness of gender, if we continually recognize that notions of gender have been so intricately interwoven in the fabric of our culture that neither historical actors nor historians have escaped their influence."[1]

This chapter has three parts. The first part provides some historical background. It considers the history of consumption, in particular the ways it has been assigned to women, and the history of production, especially the ways it has been assigned to men. How have women defined their femininity in relation to designing, making, selling, buying, and using things? How have men defined their masculinity in relation to using, buying, selling, making, and designing things? And how have the two fit together? The concept of gender, I suggest, makes sense only if we look at both sides of the story. Likewise, we must look at production and consumption together if we are to understand either. As Nina Lerman, Arwen Palmer Mohun, and Ruth Oldenziel argue in their recent article "Versatile Tools: Gender Analysis and the History of Technology," all of these relationships must be viewed in reciprocal terms.[2]

In the second part of the chapter I look at the role of gender in the way historians have defined—I shall argue, misdefined—technology. We have tended to look only at the "male" part of technology: production, not use; trains but not eggbeaters; military consulting but not cooking instruction. I argue that we can not understand technology and technological change when we accept such a narrow definition of technology. Not only the definition of technology but also the way that we think about the structures of technology must change.

The final part of the chapter is prospective and prescriptive, suggesting some areas to investigate, some ways to put this new definition to use. I very briefly consider some case studies of the borderlines between production and consumption, which suggest that the spheres of masculinity and femininity, production and consumption, are not as separate as they might seem, the boundaries not as well defined. My examples are intended to suggest some of the places historians might look if they wish to understand better the issues of gender, technology, and markets. These examples—and all of the chapters in this volume—show well what might be gained by focusing on these issues.

His and Hers

His and hers: simple words that describe enormously elaborate constellations of ideas. When we say the word *his*—that is, when we talk about masculinity—we bring to mind a related set of notions, ideologies, materials, cultures, even colors. The same occurs when we say *hers*.

These words describe gender. Gender is an ideological and cultural construction, the complex of cultural ideals and ideas tied to the words *male* and *female.* Masculinity and femininity are socially constructed and historically contingent. (Pink was once thought appropriate for boys, blue for girls.) They reflect (and reinforce) power relations in society. They change over the course of time. They are political and economic as well as cultural issues.

It is important to remember that the ideals I shall be outlining here are by no means held by everyone. People's ideas about gender, like their ideas about other aspects of culture, differ. And gender conventions do not always determine the way people behave; there are many people who by necessity or by personal inclination do not bow before this kind of cultural pressure. Still, to understand technology and gender in history we have to look at what it has meant, in general, for given groups or individuals to be masculine or feminine at given times and places.

The modern story of technology and gender begins with the industrial revolution and the related consumer revolution of the late seventeenth and eighteenth centuries. I shall examine the consumer revolution first. Over the course of the eighteenth century, first in England and then in America, there was a proliferation of goods, along with new ideas of comfort; more objects were more affordable, more available, and more desirable than ever before. By the 1830s America seemed to many observers to be a country obsessed with economic gain, consumerism, and materialism. Ann Smart Martin sums up the evidence: by the early nineteenth century, she writes, "Americans already were avid consumers." A "vast flood of new consumer goods had begun to break."[3]

When and how, and in what cultural circumstances, did the belief that buying things was women's work become widespread? Recent scholarship suggests that the strong connection between women and consumption began in the late eighteenth and early nineteenth centuries. Because the domestic sphere was women's sphere, buying and arranging domestic objects was women's work. Mary Ryan, in *Womanhood in America,* found that starting in the second half of the eighteenth century "shopping played an increasingly prominent role in personal accounts of how American women spend their days."[4]

Historians have used the concept of "separate spheres" to explain nineteenth-century changes in gender ideals. The separate-spheres argument, while it is much contested, has brought forth a flowering of discussion of "women's culture" in America. Separate spheres, wrote Michelle Rosaldo, are "an opposition between 'domestic' and 'public' that provided the basis of a structural framework necessary to identify and explore the place of male and female in psychological, cultural, social and economic aspects of human life." Linda Kerber noted that the idea of separate spheres has been used to explain "an ideology imposed on women, a culture created by women, a set of boundaries expected to be observed by women."[5]

It is this last notion, about boundaries—closely tied to the power relationships that enforce them—that I would like to use in my exploration of gender, production, and consumption. Why did the belief that domesticity was women's sphere and the related idea that the job of domesticity was the buying and arranging of objects come together as they did? Ann Douglas, in *The Feminization of American Culture,* suggested that female consumption was a result of women's being driven out of the sphere of production because of the industrial revolution. That is, to put a complex argument much too simply, women retreated to piety and consumption because they did not have much else to do, a response to a loss of political and economic power.[6]

Some historians suggest that Douglas partakes of a "productivist bias" that misses the complexities of consumption, as well as its political and cultural power. Consumption, they argue, draws on a long theological tradition of argument about the "civilizing" influence of luxury and refined objects. Lori Merish has recently argued that this change was tied to a new belief in "pious consumption," the idea that "refined domestic artifacts would civilize and socialize persons and awaken higher sentiments," especially religious sentiments.[7] Indeed, many of the new artifacts—elaborate equipment for dining, more tables and chairs, cutlery, napkins, tea equipment, card tables, wigs, jewelry—were tied to a growing interest in gentility and sociability.

Some scholars contend that this consumer revolution, based on the new ideologies of liberalism and capitalism, brought on the industrial revolution, and not the other way around. A common notion of political economy in the Jacksonian era held that the increased availability of goods would make workers work harder and the wheels of industry spin faster. Perhaps the industrial revolution was not supply-side driven by new technologies but rather demand-side driven by the desire for new goods based on new religious, cultural, and political ideas. Changing ideals of the proper work of women as consumers, then, helped drive the industrial revolution. I shall discuss this further in the next section.

When summarizing the rise of ideals that are today familiar it is easy to make them seem to have been inevitable. But the notion of consumerism as an appropriate activity for Americans, and especially for women, was part of a political battle over "refinement" in the 1830s. Some people thought refinement was antidemocratic, "putting on airs." Many people resented the new goods as facilitating class division. Jacksonians opposed to luxury were seen by their political rivals as opposing economic progress and material refinement. The belief in communal use of property, for example, was directly opposed to the belief in the importance of private property rights that was so important to the consumer revolution.[8]

The political discourse was also a moral discourse. According to Jackson Lears, "Nineteenth-century political thought politicized the discourse of authenticity, locating public virtue in plain speech and plain living, disdaining the 'parasitical' vices of commerce, celebrating the leather-aproned 'producer' as the ultimate embodiment of republican reality. The producers were invariably male, the parasites effeminate."[9] To those who favored the new ideals of consumerism it seemed that it was not just a battle of Jacksonian versus Whig, republican versus liberal, or lower class versus upper class; it was also to some extent a battle of men versus women.

This overlap of feminine refinement and upper-class refinement—of the purchase of goods with the upper classes and femininity, and the production of goods with the lower classes and masculinity—was to have its fullest expression in the new age of advertising and consumption that would begin in the late nineteenth century.

Again, there is much debate about when this second consumer revolution began and why. But the outlines of the change are clear. The rise of department stores, the vast increase in both the size and the cultural authority of the middle class, and the changing role of women were all part of this new era. Lears suggests that there was a "fundamental cultural transformation" between 1880 and 1930. He and others have pointed to the rise of advertising agencies, the changing work force, shifts in family spending, and new credit institutions as evidence of this transformation.[10]

Warren Susman has written the most evocative short description of this new era. He suggests that the whole culture after about 1890 was built on the vision of a possible world of abundance: "Key words began to show themselves: plenty, play, leisure, recreation, self-fulfillment, dreams, pleasure, immediate gratification, personality, public relations, publicity, celebrity. Everywhere there was a new emphasis on buying, spending, and consuming. Advertising became not only a new economic force essential in the regulation of prices but also a vision of the way the culture worked: the products of the culture became advertisements of the culture itself."[11]

Like the first consumer revolution, this one did not come without political struggle. "One of the fundamental conflicts of twentieth-century America," Susman writes, "is between two cultures—an older culture often loosely labeled Puritan-republican, producer-capitalist culture, and a newly emerging culture of abundance. If twentieth-century American politics rarely carries the burden of ideological conflict, there was nonetheless a significant and profound clash between different moral orders. The battle was between rival perceptions of the world, different visions of life. It was cultural and social, never merely or even centrally political."[12]

It is in advertising that we can best see signs of this second consumer revolution, and see that like the first, it was associated much more with women than with men. Roland Marchand outlined advertisers' general beliefs in *Advertising the American Dream.* In the 1920s, he writes, advertisers believed that the audience for advertising was women. The words they used to describe this audience were words they thought applied to women: *emotional, capricious, irrational, passive,* and *conformist.* Advertisers, Marchand writes, "became increasingly committed to a view of 'consumer citizens' as an emotional, feminized mass, characterized by mental lethargy, bad taste, and ignorance." Advertising trade journals assumed that more than 80 percent of consumers were women. In 1937 *McCall's* magazine restated the common belief: "Categorically . . . man is always the producer . . . woman, the consumer."[13]

This belief shaped advertising profoundly. Advertising leaders wrote that women possessed a "well-authenticated greater emotionality" and that therefore advertisements must be emotional. Women were characterized by "inarticulate longings," and therefore advertisements portrayed idealized visions, not "prosaic realities." Because women wanted personal stories, advertisements aimed at being "intimate." "We must remember," wrote one advertiser, "that most American women lead rather monotonous and humdrum lives. . . . Such women need romance. They crave glamour and color."[14] Advertisers, in short, accepted the ideology of gender and shaped their advertisements to fit it. The second consumer revolution, like the first, was gendered female.

It is easier to know what advertisers claimed than what women actually thought. Susan Porter Benson has pointed out that much of the practice, if not the ideology, of female consumption was confined to the upper and middle classes.[15] Elaine Abelson's study of middle-class shoplifters suggests that women used the excuse of irrational desires created by department stores and advertising to claim that they were unable to control their thefts. "It is not that we need so much more, or that our requirements are increased," one woman explained, "but we are not able to stand against the overwhelming temptations to buy which besiege us at every turn."[16] Whether that was an excuse or an explanation is of course uncertain. We should not assume that the point of view of those who would create culture is accepted by those who are part of it. There is always resistance and reaction, and culture is always created from the bottom up as well as from the top down. There was a continual struggle for, rather than the imposition of, a hegemonic ideology of female consumption, as Susan Smulyan has written.[17] One should not take the advertisers' words as the simple truth.

Why did advertisers devise—or accept—this scheme of production and consumption? In part, as we have seen, they were simply making use of beliefs that had been around for at least one hundred years and reshaping them for their own benefit. But Marchand suggests that it was in part a defensive technique: "The world of manliness might thus stand aloof from the threats posed by the new ethic's feminine tendencies toward frivolousness, emotionality, and effusive adornment. The idea that extravagances in consumption were the particular vice of women served as an opportunity to spare advertising men from concerns that the new ethic might plunge the entire society into an abyss of hedonism. Men might safely overindulge women if they walled off the sphere of consumption from their own sphere of work and rational control."[18]

Let me turn now to this other sphere, this "sphere of work and rational control," the sphere of industry and technology, the sphere of production. I shall ask three questions here. How was it that as consumption was being defined as women's work, most technology and industry—production—was being defined as men's work? What did this definition mean to the course of industrialization, and to changing notions of masculinity? And how did the very definition of technology change to reflect the gendered ideas that shaped it?[19]

The industrial revolution started in the eighteenth century with new technologies of iron production, textile manufacture, and metalworking, new power sources such as the steam engine, new ideas about the division of labor and managerial control and accounting, and new transportation technologies that allowed the wide distribution of a vast new array of factory-made goods. The second part of the industrial revolution started in the 1880s. It built on new managerial techniques and new technologies of steel and electricity to vastly increase the output of goods.

Although the study of the consumer revolution has been tied from the start to the study of gender, that has not been the case for the industrial revolution. Indeed, most historians of technology have ignored gender. Judy McGaw suggests that we might put gender back into the study of the American industrial revolution by asking, simply, which came first, the sexual division of labor or industrialization? This is another way of asking the question I raised earlier about whether the industrial revolution was demand-side or supply-side driven. Based on her study of industrialization, McGaw's answer is the same as that of those who have studied consumerism. She finds that the gendered division of labor came first. "Most of the activities [that made] woman's domestic work essential in industrializing America were already central to her role long before industrialization. The sexual division of labor began raising the American standard of living well before technological change in industry sought to achieve the same goal."[20]

Moreover, the sexual division of labor, based on gender ideals, played an essential role in industrialization. McGaw writes that new ideas about gender roles may well be "the most ingenious contrivance invented in industrializing America." "The very doctrine [of separate spheres] that refused to recognize women's work served admirably to supply industrializing America with cheap and essential female labor." Women's skills acquired in work in the home prepared them to perform the manual labor that industrialization required. It "made such work appear 'natural' and 'unskilled,' and provided a rationale for women's low wages." Further, the idea that "women's place is in the home . . . meant that employers could lay off female workers at will."[21]

The other side of the notion of separate spheres—that it was natural for men to work outside the home—also affected the course of industrialization. It meant that men were more interested in money and control, and less in safety, cleanliness, or time with families." McGaw points out that because the American notion of masculinity emphasized courage and bearing pain without complaint, a male worker rarely objected to the new hazards of the workplace during industrialization: "Gender ideology probably permitted, and perhaps encouraged, the development of machinery for men's workplaces without regard to its potential danger."[22]

Ideas about masculinity thus helped to shape the industrial revolution. In turn, the industrial revolution helped to redefine masculinity. New ideals of masculinity became increasingly commonplace in the mid–nineteenth century as the economy changed and men's work came to be defined by its connection with machinery. According to David Levernz, "Three basic masculine ideals were available in the mid–nineteenth century: The genteel patrician was the cultured gentleman of the old school. The artisan valued personal independence and pride in work. The aggressive self-made man was at the center of the new business culture. He was preoccupied with power and force, imposing his will upon the world out of fear of being crushed by it."[23]

The artisan and the self-made man both used technology to define their masculinity. The preoccupation with personal independence and work, with power and force, was often directed in technological ways. You were a man not only because you could hunt and fight but also because you could control nature through the use of tools and machines. Technology could be either a positive expression of masculinity, if a man controlled it, or a negative one, as when new machines made it possible to replace men with women or boys.[24]

We can see some of the ways in which men used machinery to define themselves and their role in the world by looking at literary and artistic sources. Men began to use mechanical metaphors for masculinity. They began to think of themselves as machines, both in positive and negative ways.

The overlapping of manhood and mechanism is in many ways the theme of Hermann Melville's *Moby Dick* (1851). Ronald Takaki reads *Moby Dick* as in part a story of American technology: the ship a factory, the crew, factory operatives, "a virtual miniature of industrial America." Ahab is "a machine-man," his soul of "welded iron," his "heart of wrought steel."[25] The other sailors too act like machines, especially at times of emotional stress. At the moment when Ahab's plans to pursue the white whale are revealed, for example, the sailors are described as moving "like machines" about the deck. Actions are "mechanical" when there is great emotion involved, whether the homoerotic relation between Queequeg and Ishmael or the anxiety with which Bildad takes leave of Peleg. Ahab, at the height of emotion when announcing the prize for the killing of the Great White Whale, produces "a sound so strangely muffled and inarticulate that it seemed the mechanical humming of the wheels of his vitality in him." Acting in a mechanical way is an acceptable, masculine alternative to showing emotions.[26]

Frederick Douglass, in his 1845 autobiography, writes that the northern docks, where he worked, were silent, unlike the southern ones, and that the lack of noise confirmed his sense of the laborer's "sober, yet cheerful earnestness . . . his own dignity as a man." In the 1855 revision he explains that the silent work meant that "everything went on as smoothly as the works of a well-adjusted machine," thus conflating masculine dignity with the operation of a machine. The machine's precision and quiet, smooth functioning reflected favorably on its operator, who saw himself reflected in his machinery.[27]

The language of technology was a language that carried meanings about class and gender as well as, as in Douglass's case, race. Thinking technologically could be lower class, unrefined, and male: Lori Merish points out that in Caroline Kirkland's *A New Home* (1839) a schoolteacher betrays his ignorance of refined values, of ethical and aesthetic sensibility, displaying "his insensitivity to the 'magic' of musical performance" by attending instead to its technology, "the construction of the instrument, which he thought must have taken 'a good bit o' cypherin.'"[28]

The popular belief that nature was female and that understanding and using nature—science and technology—was men's work shows up in the ways in which the female figure came to decorate sites of technological display. At the Crystal Palace Exhibition in 1851 Hiram Powers's sculpture *The Greek Slave,* a female nude, was shown not with art objects but with machinery, a juxtaposition that may have been a way of reinforcing the appropriateness of machines for men. The sculpture's location suggested that men and machines could unveil and subjugate nature in the same way that men tried to unveil and subjugate women. The use of a female nude as the central element of the frieze at the entrance to the main building of Drexel University may have the same meaning.[29]

In the late nineteenth and early twentieth centuries, with further industrialization, the positive relationship between masculinity and technology began to change. Not only were women working in factories, with machines, more than they once had but many men found themselves in white-collar jobs, away from both physical labor and machinery. Perhaps reflecting a loss of control over the course of industrialization, and, significantly, especially after the Civil War, the literary references to machinery are less positive during this period. In *The Red Badge of Courage* (1895) Stephen Crane described men at war as "machines of steel," "machine-like fools" caught in the machinery of war, part of a mechanical chain of command, thinking mechanical thoughts. But there was still a sense that masculinity is tied to technological prowess and even managerial work.[30]

Mark Twain's *A Connecticut Yankee at King Arthur's Court* (1889), the most lyrical nineteenth-century portrait of a technologist, makes a point of tying the Yankee's technological prowess (and his role in management) to traditional notions of manliness. The Yankee introduces himself thus: "Why, I could make anything a body wanted—anything in the world, it didn't make any difference what; and if there wasn't any quick newfangled way to make a thing, I could invent one—and do it as easy as rolling off a log. I became head superintendent; had a couple of thousand men under me. Well, a man like that is a man that is full of fight—that goes without saying."[31]

In the twentieth century work became increasingly industrialized. Some Americans felt that the skilled worker was "unmanned" by his loss of skills, his place threatened with replacement by boys or women, the (male) producer culture threatened by the (female) culture of abundance. Sherwood Anderson lamented men's distance from preindustrial tools, "the plow handles, the saw, the hammer, the scythe, the painter's brush, the pen, the hoe." "Man is a doer," he wrote in his *Memoirs*. "It is his nature to find strength in doing. It is what he does through things in nature, through tools and materials, that feeds his manhood and it is this manhood that is being lost."[32]

The locus of masculinity in technology shifted from the production worker to the inventor and then to the engineer. Melville conflates fatherhood and invention ("we fathers being the original inventors and patentees").[33] Recent work in the cultural history of engineering suggests that ideals of masculinity are key to the self-image of the engineer. "The engineer," says Carroll Pursell, writing about the early twentieth century, "was a manly ideal."[34] Engineering was about control over nature, with nature often described in feminine terms. It is no surprise that engineering became one of the most male-dominated professions.

Redefining Technology

Not only did cultural ideas about masculinity help shape the course of the industrial revolution and industrialization shape ideas of masculinity but the two together shaped the very definition of technology. Technology, we all too easily assume, is what men do. This narrow definition of technology has helped misdirect and confuse studies of its history. I will spend some time on this point, for it is key to the final part of this chapter, on the directions historians of technology might take to resolve some of the dichotomies I have discussed.

There have been far fewer studies of the connection between masculinity and technology than of the connection between consumption and femininity. After all, as Carroll Pursell has pointed out, "technology is so obviously masculine that it hardly seems worth making the point."[35] But the interactions between technology and cultural notions of manhood are worth some investigation, for a variety of reasons. To do so might, as Judy McGaw suggests, allow us to understand "the social shaping of those men who made technology and the manner in which their socialization narrowed their perceptions of technological choice."[36] It might improve our understanding of the speed and direction of the adoption of new technology, especially the ways in which technology was applied differently to men's work and women's work.

Feminist historians have been the most outspoken among those questioning the definition of technology. They have argued that women's technological achievements are defined as nontechnological. Ruth Oldenziel suggests that the patent system played a role in this, defining invention narrowly as part of the "paradigmatic shift toward a machine-bound interpretation of inventions." Judy McGaw has suggested that our notion of skilled and unskilled work is based in part on the idea that housework, the work that all women knew, was defined as being "natural" for women rather than skilled. The skills of household work were deemed untechnological and unskilled because they were the common knowledge of women.[37]

In the nineteenth century technology began to be sorted into hierarchical categories, in large part based on gender. These hierarchies of skill, Nina Lerman points out, are cultural products. She argues that "technological knowledge in America has been and is perceived hierarchically" and that these hierarchies depend on not just gender but also race and class. "Social boundaries of gender and race and class," she writes, "are intertwined with technological knowledge and thus with technological change."[38]

Historians' definitions of technology have tended to conflate cultural hierarchy and technological skill. The more an activity is the province of men, especially the more it is done by white men, the more we label it "skilled," the more we call it "technological." We have fallen for a culturally defined idea of skilled technology and assumed that it was somehow innate. (I am reminded of a story that my father told me about his time as an apprentice in the Philadelphia Naval Aircraft Factory in the early 1940s. Only the most senior men were allowed to run the machines that required the most skill. When the war came, though, management brought in "girls" to run them. "Skilled work" was defined as what the senior men did, which was not necessarily work that was more demanding or required more knowledge or dexterity.)

We have tended to define male medicines as true medicine, as scientific, and the tools of mainstream medicine as technologies, while defining traditional women's healing as not scientific, and without an associated technology. Agricultural work done by men we have considered technology; that done by women, a primitive technology or not technological at all. We have viewed goods destined for market as more important than those used within the household.[39] We have neglected the work in consumption, but preparing food is as much a part of the system of agriculture as planting or harvesting it. We have downplayed the skill and knowledge required by users of technology, looking at the machine and not the task, looking for complex systems on the production side, not on the consumption side.[40] We have tended to define technology as material culture, in particular as tools.

Judy McGaw has suggested a new direction for a future history of production and consumption to solve this problem: "We can cease taking the 'separate spheres' as 'logical' units of analysis. We need not accept home and work, women's activities and men's labor, as separate simply because Americans chose historically to separate them spatially and rhetorically."[41] We must consider the choices that were made and think about why those choices were made and the role gender played in them.

Mapping the Borderlands

In the final part of this chapter I want to do just that. I want to revisit the history of production and consumption with gender as the analytic tool. The case studies I shall look at suggest some of the ways in which ideas of gender and technology have been expressed in different times and places. But all of these studies have a moral—one of the main points of this chapter—namely, that the lines are harder to draw than they at first seem to be. There are always gray areas, and connections between the spheres of masculine and feminine, and it is in these gray areas and along these connections that the most interesting work is to be done. Studying these borderlands can cast light on the big questions of consumption and production in a way that studying either side on its own never can. Here I shall roughly map some of the borders between the continents of consumption and production, between his and hers.

I shall examine case studies in three areas of the relationship between technology and gender—production technologies in the feminine sphere and consuming behaviors in the masculine sphere; entertainment technologies; and mediators between production and consumption. In all of these studies users have to negotiate the usual gendered categories.

"Women's technologies," technologies used mostly by women, are one place to look for the contradictions and limitations of the simple story of production and consumption. The very term seems odd, almost an oxymoron, given the notion of separate spheres. But a quick look at some of the sites where women control technology suggests some interesting questions. The areas I shall touch on include household technologies—the machine in the parlor, so to speak—and women at work in the factory and office.

Household technologies are, generally speaking, technologies of consumption, and so they have been difficult to understand in the traditional framework. Women who work at home have been neglected or misunderstood by historians of technology. Until recently, Judy McGaw has noted, historians of technology who looked at women asked the question, "How has technological change affected women?" That is, they viewed women as passive victims of technology. However, the best general scholarship asked, How has society affected technology? How have men come to invent, develop, transfer, and use new machines and processes? They viewed men as active users of technology. She urges us to ask the same questions of women.

The first historian to ask these more interesting questions of the history of housework was Ruth Schwartz Cowan. In her *More Work for Mother* she portrayed housewives as women who made meaningful choices. "Cowan's housewives," writes McGaw, "emerge as leading contributors to the success of American technology. They are active shapers, not passive victims." They were, in other words, "not so different from other people historians of technology have studied, people such as inventors, entrepreneurs, corporate managers, and skilled workers." Moreover, housewives, and housework, are part of a larger picture. Cowan shows how household technologies are parts of technical networks of food, clothing, health care, transportation systems, water, gas, and electricity. The household, writes Cowan, is "part of a larger economic and social system; and if it did not constantly interact with this system, it could not function at all—making it no different from the manufacturing plant outside the city or the supermarket down the street."[42]

Machines intended for use in the household—women's machinery—are also easier to understand if we consider the boundaries of gender ideals that they have to cross to be adopted. Consider the sewing machine. The long and troubled history of the invention and acceptance of the sewing machine might be seen as an example of the problems of technologies that do not fit well into the dichotomy of producers and users, domestic and public, male and female. The first to invent a workable sewing machine was Barthelemy Thimonnier, between 1830 and 1841. No sooner had he set up a shop full of machines than his shop was destroyed by an "enraged mob of tailors and seamstresses." Walter Hunt, who independently reinvented the device, met a different sort of resistance. "Delegations from needlewomen's aid societies beseeched him to destroy it," J. M. Fenster writes. "Clergymen joined the chorus against Hunt's sewing machine. . . . Ministers and priests called on Walter Hunt to warn him that the sewing machine would drive thousands of displaced needlewomen to prostitution or other crimes." His daughter wrote that "the introduction of such a machine . . . would be injurious to the interests of hand-sewers. I found that the machine would be very unpopular and . . refused to use it."[43]

This story of resistance to the sewing machine can be read as a story of resistance to technology intruding into the women's sphere. Not until Isaac Singer figured out how to market the idea to women was the sewing machine accepted. He did this by decorating his shop to look "like a home, not a store, and certainly not a factory."[44] His merchandising techniques, including credit sales, installment plans, and saleswomen, would be adopted by other sellers of machines to the domestic marketplace. Singer and his followers domesticated technology by making it seem appropriate to the sphere of consumption, the sphere of women.

The telephone too had a difficult and interesting early history because its use defined—and defied—gender ideals. The inventors and early promoters of the telephone saw it as a business tool. They did not approve of the way women used telephones for their own purposes, as a social tool, part of the work of building community. From a male point of view this seemed frivolous, almost subversive, not a suitable use for the technology. It took many years for the telephone companies to change their marketing strategy and begin to encourage and exploit women's use of the phone, for example, by charging not for the connection made but for the duration of the call.[45]

From women as consumers of technology let us turn to women as producers in what traditionally has been thought of as the sphere of masculine work. Although women have worked outside the home for a long time, it has never been easy for the culture to accept this; it does not fit with our ideas about technological appropriateness. So we have had to make work somehow acceptable for women by either changing the work or changing the way we defined women's place. "For understanding woman's place in the nineteenth-century workforce," writes Judy McGaw, "identifying the links between the 'separate spheres' is crucial." Here I shall briefly examine the rhetoric of women's work at Lowell, Massachusetts, at the Philadelphia School of Design for Women, and in the introduction of new office technologies.

Lowell, Massachusetts, founded in 1821, was a planned industrial town. Young women made up more than half of the work force. These "mill girls" were perhaps the most famous innovation of the northern New England textile system. Young women from the Massachusetts and New Hampshire countryside, for the most part, they came to Lowell to work to earn cash for themselves or their families. Most were between fifteen and thirty years old—almost half were between fifteen and twenty—and most of them stayed at Lowell for only a few years. There were both economic and cultural reasons why the Lowell mills depended on women. What I want to examine here is the language the mill owners used to describe the mill girls' work, language that reflects an attempt to make industrial work appropriate to women.

The owners of the mills established a detailed set of rules that summarized the relationship between the mill girls and the mill managers. These "General Regulations" were part moral exhortation, part a guide to the mill girl's duties, and part a legal contract stating her obligations as an employee. She was required to "attend assiduously" to her duties, to "aspire to the utmost efficiency" in her work, and "to evince . . . a laudable regard for virtue [and] temperance." She would be expected to attend public worship, to observe the Sabbath, and not to drink or gamble. At work, she would "conform to regulations."

The regulations went on to outline a social contract that used the language of republicanism in the service of capital: "[Persons in the employ of the mill] will perceive that where objects are to be obtained, by the united efforts and labor of many individuals, that some must direct and many be directed. That their religious and political opinions need not however be influenced, nor their personal independence, or self respect, or conscious equality lost sight of or abandoned." Finally, the regulations made explicit the place of the mill management *in loco parentis,* not only in the legal sense but also in the moral: "[Employees] may apply with confidence to the Agent for advice; and such aid and counsel as he can afford them, will be cheerfully granted, especially to those who may be far from their parents and friends. It remains to encourage and cherish mutual respect, kindness and conciliation towards each other, and that peculiar instances of industrious and honest merit be rewarded, and which the Agent will reciprocate and aspire to accomplish."[46]

Similar language was found in the boarding house rules. Boarding house keepers, like the agents of the mills, stood *in loco parentis,* with an obligation to report "all cases of intemperance and dissolute manners" to the mill management. They were responsible for "rendering their houses comfortable, tranquil scenes of moral deportment, and mutual good will."[47]

The language would have seemed reasonable to the mill girls, for it was the language of the family. It was, especially, the language of discourse between men and women. The mill owners needed to use women as workers, and so they had to design their relationship with their workers to reflect prevailing ideas about the place of women in society.[48]

The same principle can be seen in a very different sort of women's work. Nina de Angeli Walls, who has studied the Philadelphia School of Design for Women, finds that the rhetoric of the separate spheres was adapted to make certain sorts of work acceptable for women. "Supporters of the school took pains to assert that artistic endeavors remained within woman's natural sphere, since women were believed to have an innate sensitivity to the fine arts. A widespread belief in the spiritually uplifting effect of exposure to the arts also made women seem especially suited to this field."[49]

The founder of the school, Sarah Peters, was aware of the problem of crossing the boundary dividing production and consumption and sought a rhetorical solution. She wrote in 1850: "I resolved to attempt the instruction of a class of young girls in the practice of such of the arts of design as were within my reach. I selected this department of industry, not only because it presents a wide field, as yet unoccupied by our countrymen; but also because these arts can be practiced at home, without materially interfering with the routine of domestic duty, which is the peculiar province of women."[50]

We see here a careful balancing of industry and domestic duty, of the "arts" being gendered female despite their being situated in the male sphere of production. The rhetoric of femininity allowed women to participate in the economy in ways that would otherwise seem inappropriate.

Office technologies, beginning with the typewriter, have generally been regarded as a female realm. Introduced in offices in the 1870s and 1880s, the typewriter almost immediately became a women's machine, but it played only a minor role in the feminization of office work. The growing number of educated women looking for work in the late 1800s was more important, as were managerial decisions to establish a division of labor in clerical work. Women who acquired the specialized skill needed to run the machine fit well into the new subdivided work structure. Because it was a new machine, with no tradition requiring either a male or female operator, the typewriter allowed women to come into the office without displacing men. By 1920 office work had become women's work.[51]

If office machines fell into the women's sphere, what was to be done with computers? In some ways programming computers seemed to be a clerical sort of job. Women had traditionally had the job of "computer," the person who did calculations for scientists, and so it seemed to make sense for them to take over the care and feeding of the new calculating machines. Moreover, World War II military service and other war work for men meant that women were chosen as programmers for the early computers in part because there were few men available. Many of the earliest programmers of the new computing machines were women with math and science backgrounds, often with Ph.D.'s in mathematics.

As the field expanded, men came to outnumber women. According to the historians Beth Parkhurst and Joan Richards, who analyzed the hiring of programmers in the 1950s and 1960s, few women were hired to work as business programmers. In the business world programming was a prestigious, highly paid job, and so was mostly reserved for men. Women continued to be accepted as programmers in scientific and engineering fields, though, in part because programming had a comparatively low status in those fields. Overall, in 1967, 20–30 percent of programmers were women. "Girl programmers," a 1967 guide to careers in computer programming suggested, could do every bit as well as men.[52]

There are few historical studies of men's consumption behavior to contrast with these many studies of women's work in production. The peculiar nature of automobile marketing, a topic too little studied, suggests the contradictions in marketing to men. Roland Marchand has called attention to Henry Ford's "resistance . . . to the luxury-minded softness of the consumption ethic." Ford insisted that the car was just a technology: "efficient, satisfactory transportation." But it was also a consumer product and had to be redesigned and sold in new ways to take into account the gender encoding of sales. That became evident in streamlined design, in new advertising techniques, and also, perhaps, in the "highly unethical" sales practices that came to characterize the industry.[53]

Entertainment technologies, the second of my three categories, are technologies of neither production nor consumption but partake of both, occupying an in-between ground. Success in the movie or recording industry requires not only a well-tuned eye and ear for consumer tastes but also, and this was true especially in the early days, a profound knowledge of the technologies. Thomas Edison certainly had technological ability, but he tended not to understand the consumer side of the equation. Edison, I shall argue, failed to convert from a producer ethos to a consumer ethos at the time of the second consumer revolution. An examination of the changing nature of American markets can help us understand Edison's business problems.

Bernard Carlson, working from the theoretical structure of the "social construction of technology," has suggested that in any culture a technology might be used in many ways. Inventors succeed by finding solutions to both technical and cultural problems. In his terms, inventors "make assumptions about who will use a technology and the meanings users might assign it. These assumptions constitute a frame of meaning inventors and entrepreneurs use to guide their efforts at designing, manufacturing, and marketing their technological artifacts."[54] Edison, he suggests, made the wrong assumptions.

Edison sold his first movie machines to arcades, where they found a short-term popularity. But Edison was not interested in entertainment films, for the most part, and certainly showed little interest in giving the public what it wanted. He lost out on the mass market that developed in the early twentieth century, focusing first on patent battles and then on the sorts of films that appealed to him and what he thought of as middle-class tastes. Other filmmakers made popular romances and films that were modern in that they featured sex and violence. Motion pictures were part of a new consumer culture that included celebrity, pleasure, and leisure, and Edison wanted no part of it. Rather than to stars and popularity he turned to better technology, educational films—the films he thought people should watch rather than what they wanted. Edison went out of the movie business in 1918.

Edison had similar problems with the phonograph. He thought of the phonograph as a business machine, for taking telephone messages, for dictation, and even thought that cylinders instead of letters might be sent through the mail. He was opposed to the use of the machine for "amusement purposes." "It is not a toy," he wrote.[55] But of course the consumer market for the phonograph was as a toy. Edison never understood this. He cared more about perfect reproduction and "good music" than selling records and thought that the record manufacturer's job was to educate the user about music, not to sell the records the customer wanted. He refused to record jazz because he did not like it and was convinced that his favorite song, "I'll Take You Home Again Kathleen," was the most popular recording.

Carlson points out that these technologies were invented just as American culture was switching from a producer culture to a consumer one. Edison missed the change. He was without question a product of the culture of production. He looked to business markets; he did not believe in advertising; and he certainly did not believe in mass markets. He designed his motion-picture machines, and marketed them, in much the same way that he designed capital goods such as the electric-power system.

Where we see a conflict between consumption and production we should also consider the way gender enters the story. Edison and many of his contemporaries saw this sort of entertainment as womanly. "These zeotropic devices [i.e., motion pictures] are of too sentimental a value to get the public to invest in," he wrote.[56] In Edison's day, according to a period dictionary, *sentimental* meant "addicted to indulgence in superficial emotion" or "apt to be swayed by sentiment," words Edison might well have used to describe women. And indeed the movies, like recorded sound, were thought to have special appeal to women. The entertainment industry depended to a large extent on women as purchasers and users and partook of a female ethos of consumption. Edison was uncomfortable working in that world.

The radio too caused problems because of the way it crossed gender boundaries. Like most machines, the radio was originally a masculine technology, an ugly box that sat in the garage or attic, where men could prove their technological prowess by finding distant stations. But as Susan Smulyan discovered, this had to change when radio became commercialized. The market for radio commercials was, of course, women—remember, consumers were defined as women—but the radio audience was thought to be male. And so the design of the radio had to be changed to make it look more like furniture so that it would fit in the home.

Not just the receiver but also the programming—especially daytime programming, addressed to women—needed to be changed in order to allow "advertisers to sell and at the same time present their product in a warm and friendly atmosphere." The answer was to play up radio as a visitor to the home and to play up the idea of the home and family as the site of consumption.

Those promoting radio advertising first had to overcome resistance to advertising "from every group interested in early radio: listeners, educators, critics, legislators, and regulators." They then had to overcome resistance to radio advertising to women, which they did by turning radio into an instructional tool staffed by home economists. And so in the mid-1920s there appeared shows about how to cook, how to shop, how to make clothing, and how to do housework. Educational programs, advertisers thought, could more easily influence purchasers than entertainment programming. Soap operas, which became the most successful radio shows for women, followed the criteria that the instructional shows had developed: they were personalized, short, sold to companies that made a product that women bought without consultation—soap. Radio was transformed from an evening family entertainment controlled by the father to a women's medium. "Moreover," Smulyan writes, "the introduction of commercialized broadcast radio meant that the definition of the consumer as female had to be fought over again, readjusted, and even expanded." Daytime radio in turn "helped reinforce the ideology [that] united women with consumption."[57]

My third set of stories highlight some of the people who worked on the borderline between production and consumption. Starting in the early twentieth century a new kind of mediator arose in the corporate world, individuals—often women—whose job was to ensure that producers and consumers were on the same wavelength. They worked on both sides of the producer-consumer divide, assisting producers to design products that would sell and assisting purchasers to find out about and use the new products. These individuals went by many names—advertising agent, home economist, product designer. Smulyan suggested that the rise of these mediating professions represents as profound a change in the corporate world as did the rise of middle managers in the nineteenth century.

According to Carolyn Goldstein, the home economists of the 1920s saw themselves as mediators between producers and those who would buy their products. Where once home economists had spent their time helping with home production, now they helped with home consumption. Economics, the *Journal of Home Economics* announced in 1927, was "the science of ultimate consumption." Home economists advised housewives on how to use products and advised manufacturers about those markets.[58]

They found a ready audience in both camps. Producers needed someone to listen to and communicate with consumers, especially the women they thought represented 85 percent of the market. Consumers, overwhelmed by the flood of new products, turned to the home economists for advice and help.[59] Goldstein has shown how this group of women affected not just production in the home and consumption but also the wide range of new technologies that went into the home. Refrigerators, just becoming popular in the late 1920s, presented a sales problem for their makers and a use problem for those who purchased them. Manufacturers needed advice on what was wanted and needed, and housewives needed advice on how to use them. Home economists filled this role admirably, mediating between the desires of the manufacturers and those of the consumers. They researched the use of foods, the proper temperatures for foods, the amounts of space needed, the costs of refrigerators, as well as more technical aspects of the alternative technologies of refrigeration and conducted polls to discover attitudes toward refrigeration. They passed this information on to the manufacturers and electricity utilities as well as to housewives through government pamphlets and women's magazines. They even devised new cookbooks to take advantage of refrigerators and updated old recipes.

Home economists saw production and consumption not as separate processes but as two sides of a coin. That metaphor is too simple, though, because the linkages between production and consumption were in fact quite complex. Home economists helped both sides in the negotiations of consumption and production. As Ruth Oldenziel has pointed out, home economists took an active part in the creation of technical artifacts: "These women professionals not only mediated, but also helped frame the meanings of technical products. They were therefore participants in and producers of a culture of consumption."[60]

While home economists educated and changed users and producers so that their interests might fit more closely, advertisers educated consumers to convince them to buy. Advertising agents saw their job as educating the public to understand what products were available, how to use them, and, most important, why they needed them. Advertisers saw themselves as indispensable go-betweens, serving both producer and consumer. The served the producer by increasing the size of the market. They served the consumer, they liked to claim, by educating her.

Although the vast proliferation of products and brands was seen as a sign of advancing American civilization, it posed problems for what advertisers saw as the poor, confused woman consumer, with her "more or less chaotic condition of mind." Advertising viewed itself as a therapeutic profession whose job it was to help the bewildered consumer, who had the "task of sifting and choosing just those things they really want, just those things in harmony with each other." Especially in the case of technological items, advertising made it possible for a woman to understand what was available to her. The poor woman in a Parke, Davis and Company advertisement had this problem: "I used to be all confused. I would go into the drug store and see rows and rows of products, all different. I suppose most of them were good—but how could I know? . . . I almost wore my mind out trying to decide."[61]

Since the poor woman consumer could not be expected to know enough about the new technology-based products, the advertising firms took on the job of educating them. Widely held ideas about gender roles meant that men would know about these things, and so advertisers frequently featured male experts in their advertisements. Roland Marchand highlights the gender implications of advertisers' use of experts: "The inadequacy ascribed to women was only one salient example of the wider public incompetence that advertisers assumed, and sought to reinforce, by their constant celebration of experts."[62] Advertisers, along with the experts they brought to the magazine page or radio program, attempted to bridge the gap between producers and consumers, passing on to consumers the specialized knowledge they needed to understand the brave new world of consumption.

Advertisers tried to educate and convince consumers. Buyers, marketers, and designers worked in the other direction; their job was to design products that consumers would want to buy. Buyers kept track of what sold, what consumers seemed interested in; they kept statistics and watched consumers at work. Designers of consumer products visited stores to see what seemed popular, took into account past sales and popular trends, and tried to produce the goods consumers wanted to buy. Surveys, customer observation, and market research brought about an early-twentieth-century revolution in marketing, allowing the producer to target the customer more closely than ever before.

In her study of Frederick Rhead, at the Homer Laughlin China Company, Regina Blaszczyk discovered an enormously sophisticated system of "accommodation" to the customer. Rhead, who was the art director at Homer Laughlin, had a complex theory of market segmentation that meshed with information and demands from buyers for department stores, direct observations of customers, and other data about consumer preference. He considered himself a trained observer of consumption patterns and saw his job as matching the technology of production with the demands of consumption. In his view, production and consumption worked together, and people like him were to guide them. Based on her studies of Rhead and others, Blaszczyk writes that "the development of household accessories for the expanding twentieth-century American consumer culture was a complicated, interactive procedure, not an autonomous artistic endeavor or marketing endeavor."[63]

Conclusion

The separate spheres of masculine and feminine, production and consumption, turn out not to be so separate after all. There is a constant interplay between them, and a wide area of interaction and connection. This does not mean that the notion of separate spheres is not useful for understanding the past. By looking at where the spheres intersect one another we can understand how both sides worked. Historians of technology have tended to draw too clear a distinction between producers and consumers, assuming that the producers alone are the technologists. The case studies examined here suggest that we should instead look at technology as a negotiation between producers and consumers, makers and users.

Many of the chapters in this volume do just that. As you read them, keep in mind the ways in which creation of consumer society is an ongoing process. Technological objects are a negotiation of value and meaning between manufacturer and worker, consumer objects a negotiation between producer and buyer. Each person who participates in the design, manufacture, sale, or use of an object brings meaning to it; each helps construct it. The ideals of masculinity and femininity, themselves shaped and constructed by the objects and actions that men and women design, make, sell, buy, and use everyday, are a key part of those meanings.

Notes

1. Judith A. McGaw, "No Passive Victims, No Separate Spheres: A Feminist Perspective on Technology's History," in *Context: History and the History of Technology: Essays in Honor of Melvin Kranzberg,* ed. Stephen H. Cutcliffe and Robert C. Post (Bethlehem PA: Lehigh Univ. Press, 1989), 173.

2. Nina E. Lerman, Arwen Palmer Mohun, and Ruth Oldenziel, "Versatile Tools: Gender Analysis and the History of Technology," *Technology and Culture* 38 (Jan. 1997): 5.

3. This summary is based on Ann Smart Martin's overview, "Makers, Buyers, and Users: Consumerism as a Material Culture Framework," *Winterthur Portfolio* 28 (1993): 141–57. For a more broad-ranging theoretical overview see Victoria de Grazia, "Changing Consumption Regimes," in *The Sex of Things: Gender and Consumption in Historical Perspective,* ed. Victoria de Grazia (Berkeley: Univ. of California Press, 1996), 11–24.

4. Mary P. Ryan, *Womanhood in America, from Colonial Times to the Present* (New York: New Viewpoints, 1975), 50, quoted in Lori Merish, "'Hand of Refined Taste' on the Frontier Landscape: Caroline Kirkland's *A New Home, Who'll Follow?* and the Feminization of American Consumerism," *American Quarterly* 45 (1993): 515n.

5. Michelle Rosaldo, "Woman, Culture, and Society: A Theoretical Overview," in *Woman, Culture, and Society,* ed. Michelle Rosaldo and Louise Lamphere (Stanford: Stanford Univ. Press, 1974), 23; Linda Kerber, "Separate Spheres, Female Worlds, Woman's Place: The Rhetoric of Women's History," *Journal of American History* 75 (June 1988): 17. For the debate over the notion of separate spheres see Dorothy O. Helly and Susan M. Reverby, eds., *Gendered Domains: Rethinking Public and Private in Women's History* (Ithaca: Cornell Univ. Press, 1992). The best summary of the notion of separate spheres in the business context is Angel Kwolek-Folland, *Engendering Business: Men and Women in the Corporate Office, 1870–1930* (Baltimore: Johns Hopkins Univ. Press, 1994), 9–11, and, for bibliography, 247–48. For the notion in the labor context see Ava Baron, "Gender and Labor History: Learning from the Past, Looking to the Future," in *Work Engendered: Toward a New History of American Labor,* ed. Ava Baron (Ithaca: Cornell Univ. Press, 1991), 12–13. For the use of the idea in the history of technology see Nina E. Lerman, Arwen Palmer Mohun, and Ruth Oldenziel, "The Shoulders We Stand On and the View from Here: Historiography and Directions for Research," *Technology and Culture* 38 (Jan. 1997): 25–30.

6. Ann Douglas, *The Feminization of American Culture* (New York: Doubleday, Anchor, 1988); see esp. ch. 2, "Feminine Disestablishment," 44–79, for an in-depth discussion of this transformation in female economic, social, and political roles.

7. Merish, "Hand of Refined Taste," 487.

8. For the politics of refinement see Richard L. Bushman, *The Refinement of America: Persons, Houses, Cities* (New York: Alfred A. Knopf, 1992; Vintage, 1993), 409–13, 425–31.

9. T. J. Jackson Lears, "Sherwood Anderson: Looking for the White Spot," in *The Power of Culture: Critical Essays in American Culture* (Chicago: Univ. of Chicago Press, 1993), 15.

10. T. J. Jackson Lears, "From Salvation to Self-Realization: Advertising and the Therapeutic Roots of Consumer Culture," in *The Culture of Consumption: Critical Essays in American History, 1880–1980,* ed. T. J. Jackson Lears and Richard W. Fox (New York: Pantheon, 1983), 3. See also William Leach, *Land of Desire: Merchants, Power, and the Rise of a New American Culture* (New York: Pantheon, 1993).

11. Warren Susman, *Culture as History: The Transformation of American Society in the Twentieth Century* (New York: Pantheon, 1984), xxiv.

12. Ibid., xx.

13. Roland Marchand, *Advertising the American Dream: Making Way for Modernity, 1920–1940* (Berkeley: Univ. of California Press, 1985), 69, 162. On the idea of the female consumer see Michael Schudson, *Advertising, the Uneasy Persuasion; Its Dubious Impact on American Society* (New York: Basic Books, 1984), 178–208; and Stuart Ewen, *Captains of Consciousness: Advertising and the Social Roots of the Consumer Culture* (New York: McGraw-Hill, 1976), 159–76.

14. Quoted in Marchand, *Advertising the American Dream*, 66, 67.

15. Susan Porter Benson, "Living on the Margin," in de Grazia, *The Sex of Things*.

16. Elaine Abelson, *When Ladies Go A-Thieving: Middle-Class Shoplifters in the Victorian Department Store* (New York: Oxford Univ. Press, 1989), 6; the quotation is from M. Jeune, "The Ethics of Shopping," *Fortnightly Review*, 1 Jan. 1895, 124.

17. Susan Smulyan, *Selling Radio: The Commercialization of Radio Broadcasting, 1920–1934* (Washington DC: Smithsonian Institution Press, 1994), 4.

18. Marchand, *Advertising the American Dream*, 162. For further discussion of how advertisers defined and manipulated the gender ideology of consumption see Stuart Ewen and Elizabeth Ewen, *Channels of Desire: Mass Images and the Shaping of American Consciousness* (New York: McGraw-Hill, 1982).

19. This is not to suggest that women did not participate in production. Rather, women's work was either rhetorically excluded and not considered economically important (as when it took place in the home) or placed into carefully negotiated, narrowly defined circumstances that isolated it from the mainstream of industrial work (e.g., the "women's work" of the Lowell mill girls). See the section entitled "Mapping the Borderlands," below.

20. McGaw, "No Passive Victims," 185.

21. Ibid.; Judith A. McGaw, *Most Wonderful Machine: Mechanization and Social Change in Berkshire Paper Making, 1801–1885* (Princeton: Princeton Univ. Press, 1987), 373.

22. McGaw, "No Passive Victims," 177. The argument could be extended to include race as well as gender; see David Roediger, *Wages of Whiteness: Race and the Making of the American Working Class* (London: Verso, 1991).

23. David Leverenz, *Manhood in the Making: Cultural Concepts of Masculinity* (New Haven: 1990), 11. See David Kuchta, "The Making of the Self-Made Man: Class, Clothing and English Masculinity, 1688–1832," in de Grazia, *The Sex of Things*, for the ways in which men's clothing reflected the masculine values of "industry and economy."

24. Masculinity was defined not only in opposition to the work that women did but also in relation to the work that boys did (see Ava Baron, "An 'Other' Side of Gender Antagonism at Work: Men, Boys, and the Remasculinization of Printers' Work, 1830–1920," in Baron, *Work Engendered*, 47–69).

25. Ronald Takaki, *Iron Cages: Race and Culture in Nineteenth-Century America* (New York: Oxford Univ. Press, 1990), 282–86. See also Leo Marx, *The Machine in the Garden: Technology and the Pastoral Ideal in America* (New York: Oxford Univ. Press, 1964), 287–319.

26. Hermann Melville, *Moby Dick* (1851; reprint, New York: Bantam, 1987), 485, 510, 57, 104, 154. In "The Paradise of Bachelors and the Tartarus of Maids" Melville used slightly different imagery to portray female paper mill workers: "The girls did

not so much seem accessory wheels to the general machinery as mere cogs to the wheels." For an analysis see McGaw, *Most Wonderful Machine*, 335–37.

27. Frederick Douglass, *Narrative of the Life of Frederick Douglass, An American Slave, Written by Himself* (1845: reprint, Boston: Beaford Books of St. Martin's Press, 1993), 102; idem, *My Bondage and My Freedom* (New York: Miller, Orton & Mulligan, 1855), 345.

28. Quoted in Merish, "Hand of Refined Taste," 499.

29. "Engines of Change: The American Industrial Revolution, 1790–1860," permanent exhibition at the National Museum of American History, Smithsonian Institution, Washington DC. See also Abigail Solomon-Godeau, "The Other Side of Venus: The Visual Economy of Feminine Display," in de Grazia, *The Sex of Things*, esp. 126; and Ludmilla Jordanova, *Sexual Visions: Images of Gender in Science and Medicine between the Eighteenth and Twentieth Centuries* (Madison: Univ. of Wisconsin Press, 1989), ch. 5.

30. Stephen Crane, *The Red Badge of Courage* (1895; reprint, New York: Bantam, 1983), 9, 41.

31. Mark Twain, *A Connecticut Yankee at King Arthur's Court* (1889; reprint, London: Penguin, 1986), 36.

32. Sherwood Anderson, *Sherwood Anderson's Memoirs: A Critical Edition*, ed. Ray Lewis White (Chapel Hill: Univ. of North Carolina Press, 1969), 387, quoted in Lears, "Sherwood Anderson," 28.

33. Melville, *Moby Dick*, 150.

34. Carroll Pursell, "The Construction of Masculinity and Technology," *Polhem* 11 (1993): 206. See also Ruth Oldenziel, "Gender and the Meanings of Technology: Engineering in the United States, 1880–1945" (Ph.D. diss., Yale University, 1992), 6; and Bruce Sinclair, "Inventing a Genteel Tradition: MIT Crosses the River, in *New Perspectives on Technology and American Culture* (Philadelphia: American Philosophical Society, 1986).

35. Pursell, "Construction of Masculinity and Technology," 206–19.

36. McGaw, "No Passive Victims," 176.

37. Oldenziel, "Gender and the Meanings of Technology," 41; McGaw, "No Passive Victims," 180.

38. Nina E. Lerman, "'Preparing for the Duties and Practical Business of Life': Technological Knowledge and Social Structure in Mid-Nineteenth-Century Philadelphia," *Technology and Culture* 38 (Jan. 1997): 36, 58.

39. McGaw, "No Passive Victims," 179.

40. Joy Parr, "What Makes Washday Less Blue? Gender, Nation, and Technology Choice in Postwar Canada," *Technology and Culture* 38 (Jan. 1997): 183–84.

41. McGaw, "No Passive Victims," 178.

42. Ibid., 175. Ruth Schwartz Cowan, *More Work for Mother: The Ironies of*

Household Technology from the Open Hearth to the Microwave (New York: Basic Books, 1983).

43. J. M. Fenster, "Seam Stresses," *American Heritage of Invention and Technology* 9 (winter 1994): 43. For a sophisticated analysis of the introduction of other machines into the clothes-making business see Wendy Gamber, "'Reduced to Science': Gender, Technology and Power in the American Dressmaking Trade, 1860–1910," *Technology and Culture* 36 (July 1995): 455–82.

44. Fenster, "Seam Stresses," 44.

45. See Lana F. Rakow, *Gender on the Line: Women, the Telephone, and Community Life* (Urbana: Univ. of Illinois Press, 1992); and Claude Fischer, *America Calling: A Social History of the Telephone to 1940* (Berkeley: Univ. of California Press, 1992), 231–36. The story of the feminization of the automobile is similar: see Virginia Scharff, *Taking the Wheel: Women and the Coming of the Motor Age* (New York: Free Press, 1991), 111–33.

46. "General Regulations, to be observed by persons employed by the Lawrence Manufacturing Company, in Lowell," 21 May 1833, Kress Collection, Baker Library, Harvard University, reprinted in Steve Dunwell, *Run of the Mill: A Pictorial Narrative of the Expansion, Dominion, Decline, and Enduring Impact of the New England Textile Industry* (Boston: David R. Godine, 1978), 44.

47. Thomas Dublin, *Women at Work: The Transformation of Work and Community in Lowell, Massachusetts, 1826–1860* (New York: Columbia Univ. Press, 1979), 75–85.

48. Ibid., 122–23. Even the architecture of Lowell reflected this gendered discourse: see John Coolidge, *Mill and Mansion: Architecture and Society in Lowell, Massachusetts, 1820–1865* (1942; reprint, Amherst: Univ. of Massachusetts Press, 1993), 160.

49. Nina de Angeli Walls, "Art and Industry in Philadelphia: Origins of the Philadelphia School of Design for Women, 1848 to 1876," *Pennsylvania Magazine of History and Biography* 117, no. 3 (1993).

50. Quoted in ibid., 184.

51. See Margery Davies, *Woman's Place Is at the Typewriter: Office Work and Office Workers, 1870–1930* (Philadelphia: Temple Univ. Press, 1982).

52. Beth Parkhurst and Joan Richards, "Women in Computer Programming," unpublished paper.

53. Marchand, *Advertising the American Dream,* 156–58; Ruth Oldenziel, "Boys and Their Toys: The Fisher Body Craftsman's Guild, 1930–1968, and the Making of a Male Technical Domain," *Technology and Culture* (Jan. 1997): 94–96; James J. Flink, *The Automobile Age* (Cambridge: MIT Press, 1990), 281.

54. W. Bernard Carlson, "Artifacts and Frames of Meaning: The Cultural Construction of Motion Pictures," in *Shaping Technology—Building Society: Studies*

in Sociotechnical Change, ed. Wiebe E. Bijker and John Law (Cambridge: MIT Press, 1992), 177.

55. Oliver Read and Walter L. Welch, *From Tin Foil to Stereo: Evolution of the Phonograph* (Indianapolis: H. W. Sams, 1976), 55.

56. Thomas Alva Edison to Eadweard Muybridge, 21 Feb. 1894, quoted in Robert E. Conot, *A Streak of Luck* (New York: Seaview, 1978), 400.

57. Susan Smulyan, "Radio Advertising to Women in Twenties America: "A Latchkey to Every Home," *Historical Journal of Film, Radio and Television* 13 (1993): 300, 311.

58. Carolyn M. Goldstein, "Mediating Consumption: Home Economics and American Consumers, 1900–1940" (Ph.D. diss., University of Delaware, 1994), ch. 1; the quotation is from 17. See also idem, "From Service to Sales: Home Economics in Light and Power, 1920–1940," *Technology and Culture* 38 (Jan. 1997): 121–52.

59. Of course consumers did not always take the advice of home economists and marketers (see Parr, "What Makes Washday Less Blue?" 153–86).

60. Ruth Oldenziel, "Object/ions: Technology, Culture, and Gender,' in *Learning from Things: Method and Theory of Material Culture Studies,* ed. W. David Kingery (Washington DC: Smithsonian Institution Press, 1996), 62–63.

61. Marchand, *Advertising the American Dream,* 343.

62. Ibid., 351.

63. Regina Blaszczyk, "Imagining Consumers: Manufacturers and Markets in Ceramics and Glass, 1865–1965" (Ph.D. diss., University of Delaware, 1995), 381.

Two

A House Divided
The Culture of the American Luxury Hotel, 1825–1860

Molly W. Berger

When Lincoln delivered his house-divided speech on 16 June 1858, he made apt use of the biblical metaphor to identify the grave crisis threatening the Union. But as the stability of the country crumbled, another sort of house—the commercial luxury hotel—prevailed as an urban monument to an idealized American society. By the time of Lincoln's oration large luxury hotels could be found in most major urban areas. These granite and marble landmarks stood as stable icons of democracy, commerce, and progress, in ironic contrast to the political turmoil of the antebellum years.[1]

Urban luxury hotels served their cities as important symbols of commercial, technological, and cosmopolitan success in ways incomprehensible to those of us jaded by the ubiquity of contemporary hotel and motel chains. While they incorporated traditional forms of architectural and decorative extravagance, these hotels also employed a new opulence of *technological luxury* to seduce consumers. The introduction of technological comforts not only created measurable standards for competition among hotels but also further complicated gendered relations within them. In these palaces of ease and hospitality technology both represented the masculine economic world and helped to define the feminized domain of domestic comfort. Technological luxury thus challenged and softened traditionally defined boundaries between production and consumption and between the public world of the economic market and the privacy of the home. The hotel's technological luxury exemplified the ironic tensions of the modern gendered world whereby the pursuit of leisured gentility and consumption was made possible by its exact opposite: energetic capitalist development characterized by work, thrift, and production.

The nineteenth-century luxury hotel illustrates how the categories of gender, consumption, and technology function as contingent parts of a larger historical process. Placed in dynamic relationship to each other, these concepts illuminate how nineteenth-century Americans adapted the built environment to the social, cultural, and economic conventions of their lives. Through the design and use of space urban luxury hotels incorporated both production and consumption and codified the hierarchies embedded in class and gender relations, public and private behavior, and the adoption of new technologies.

My argument in this chapter proceeds in four stages as follows: First, the urban commercial luxury hotel developed as a masculine technological artifact, constructed both literally and figuratively by men seeking to apply the ideology of progress to their commercial and civic enterprises. Second, the building's spatial arrangements fortified male dominance by physically grounding the abstract ideology of separate spheres in the organization and deployment of space. The hotel accommodated women and children but did so in a way that reinforced the unequal power relations between the sexes. Third, the presence of women in the hotel amid luxurious accommodations introduced a feminine, indeed potentially emasculating element to the hotel environment. At the same time that the hotel's sumptuous decoration made the hotel a more welcoming place for women, it also invoked conflicting attitudes toward luxury among men, heightening their need to clearly define male space. Finally, as a regulated and commercial public space, the hotel became a gateway to the city and to public life for women, whose prescriptive role was to remain at home. As a staging area for participation in the burgeoning consumer market, the hotel also provided feminine space where young marriageable women became commodities themselves in a marriage market whose ultimate objective was the solidification and reproduction of bourgeois class values.

The genre of hotels discussed in this chapter are urban, commercial luxury hotels built in the United States between the years 1825 and 1860.

Taverns and inns, with their rich, almost folkloric history, were precursors to the large city hotel. The word *hotel* came from the French, for whom it referred to the city homes of the French nobility. It began to be used in the United States toward the end of the eighteenth century to describe the elite group of taverns and city inns that catered to upper-class patrons. Many of these establishments were the former mansions of wealthy citizens who had moved elsewhere. Proprietors exploited the stylish architecture and decoration of these homes as well as the cachet of the departed family to court a genteel local clientele who often frequented the hotels' public rooms to attend dancing assemblies, balls, and grand political dinners. By the 1820s the large cities on the eastern seaboard all had hotels whose elegant decor, size, and elaborate services earned them the designation "first-class." An oft quoted editorial in the *National Intelligencer* of 18 June 1827 entitled "Our Public Hotels" labeled these rich buildings "palaces of the public," a phrase that has endured and particularly appeals to historians, who seem captivated by the juxtaposition of luxury and democracy and the distinctively American ability to reconcile the two.[2]

Luxury hotels developed within a context of far-reaching changes in American society that began to take hold in the first quarter of the nineteenth century. These included accelerating technological change, the onset of industrial capitalism, the increasing separation of income-producing work (especially men's) from home, urban growth, improvements in transportation that facilitated travel, a redefined concept of luxury, and the earnest pursuit of gentility by the upper and upper middle classes. As America's industrial revolution gained momentum, commerce established itself as an important component of the young nation's political economy. Many economic and political theorists believed that the manufacturing and marketing of luxury goods would contribute to the nation's economic health as the country began its transformation from an agriculture-based society to a modern industrial one. First-class hotels such as Barnum's City Hotel in Baltimore, Boston's Tremont House, New York's Astor House, the St. Charles in New Orleans, and Philadelphia's Continental not only served the wealthy classes of business and pleasure travelers but also, in responding to the marketplace demands for this type of accommodation, answered the city's need for an extraordinary, larger-than-life expression of its success, vitality, and cultural life.

Hotels evolved alongside changes in technology, financial organization, and management practices of American business in the early nineteenth century. Jefferson Williamson, in his chatty but durable 1930 history of American hotels, stated rather boldly that Boston's Tremont House, built in 1829, "was indisputably the first definitely recognized example of the modern first-class hotel." This strong statement rests on the word *modern,* for several hotels, such as Barnum's in Baltimore or the National Hotel in Washington, D.C., easily could be described as first-class. Modern hotels, however, took on characteristics of the changing world around them. Because of the escalation of size, decoration, and technological systems, a modern hotel required financing in the form of a modern corporation, rather than joint-stock investments or private partnerships. This introduced the principle of limited liability, which encouraged investment and facilitated the mobilization of capital but also limited personal risk. A modern hotel further separated ownership from management. Typically, the hotel proprietor leased the building and furnished it himself (proprietors of large hotels were always men), retaining ownership of the furnishings. In line with the growing tendency toward professionalization, investors hired professional architects to design buildings specially for hotel purposes instead of converting buildings to hotel use. Most important, developers of the modern hotel embraced the ideology of progress through their single-minded incorporation of the latest, most extensive technological systems that money could buy. More than anything else, the public profile of the hotel's technological systems projected an image of strength, innovation, wealth, masculinity, and modernity.[3]

Describing the hotel in terms of masculinity may seem paradoxical given the domestic nature of the hotel and the traditional, refined decoration, which together contribute to a feminized reading of the building. However, at the most fundamental level the hotel, owned by a corporation chartered by the state, derived its existence from the consent of its franchised citizens, specifically its men, who then imbued it with public, masculine ownership. Moreover, hotel companies enjoyed the financial support of a city's commercial elite, blurring distinctions between personal financial and civic goals but nonetheless reinforcing upper-class male ownership. The refined setting and social conventions that it implied depended on the active relationship between this world of leisure and consumption and the capitalist enterprises that produced it. As the historian Richard Bushman argues, "Capitalism and gentility were allies in forming the modern economy." The hotel world was a world in which wealthy capitalists felt at home. On the one hand, it served its community as a public symbol for the achievement of a refined, cultured civilization; on the other hand, it portrayed commercial success and an enthusiastic immersion in what contemporaries often termed the "go-ahead age."[4]

Williamson was right to begin the history of American luxury hotels with the completion of Boston's Tremont House because this establishment stood as a model of hotel design for the nation during the antebellum years. Published records of travelers to Boston during the 1830s attest to the building's novelty and its reputation as "one of the proudest achievements of American genius." The hotel project responded to a perceived need for appropriate accommodations for the many wealthy business, southern, and foreign travelers to Boston, which at the time was the nation's center for banking, shipping, and textiles. William Havard Eliot, a young lawyer who belonged to one of Boston's most prominent families, shepherded the financing, design, and construction of the Tremont through to its completion, enlisting the backing of nearly all of that city's upper class. The Boston elite invested in many large-scale initiatives, such as bridges, wharves, canals, and early railroads, which, in aiding transportation and trade, served both public and private interests. The hotel's importance transcended its value as an investment and an urban necessity, acquiring civic force through its evocation of the personal, business, and civic ethos of a class of men.[5]

The Tremont House stood at the corner of Beacon and Tremont Streets, adjacent to the Granary Burying Ground, immediately to its south. Situated in a neighborhood of upper-class homes and important community buildings, the Tremont represented and emanated the social status and power of its sponsors. The hotel became an extension of the elite's homes, reinforcing class-bound spatial and territorial boundaries, housing their guests and business associates, and serving as the venue for public dinners, celebrations, and casual associations. Isaiah Rogers, one of Boston's three pioneers in Greek Revival architecture, designed the hotel, choosing a classical Greek form: a 205-foot frontage punctuated by an impressive Doric portico supported by four massive, twenty-ton granite columns. The monumental proportions of the hotel created a physical presence in the city unequaled in any other building save the State House and the Quincy Market.[6]

It is important to see the technological side of the hotel because that is exactly what the proprietors wanted their worldwide public to see. Machines and technology conveyed masculine prestige and expressed power through their relationship with innovative capitalist development. This technological side and the commercial activities of the hotel placed it in opposition to the home, whose domestic ideology created a haven and antidote against the morally corruptive influences of the market. Observers described the luxury hotel as a "finely adjusted" or "quietly regulated machine." Hotel proprietors constructed an identity for their buildings based on technology. They found it necessary to describe these systems to the public in great detail since much of the technology remained hidden from view. Guests were aware of the plumbing, wires, heating ducts, steam engines, laundries, and kitchens only as invisible abstractions except at the point of consumption—when they turned a faucet, sat down to dinner, or eased themselves next to a radiator for a bit more heat. The technology of the building represented all that made the hotel modern, fitting it into the ideology of progress. Proponents of this ideology believed that mechanical invention and scientific knowledge provided the means to achieve a perfect civilization.[7]

Innovations at the Tremont ushered in a full century of technological competition within the hotel industry. Much about the hotel suggested its innovative nature. The huge blocks of granite that sheathed the front facade of the building and formed the columns had been transported from Quincy, Massachusetts, nine miles away, on the newly built Granite Railway, purported to be the first railway in the country. The design of the building introduced the idea of a central lobby from which radiated rooms having specific functions, such as receiving, baggage handling, and accounting. Formerly proprietors had conducted all hotel business over the bar. Single and double bedrooms with individually keyed patent locks afforded guests heretofore unavailable privacy and security. The Tremont eliminated the central-court stable, sparing guests noise from clattering hooves as well as unpleasant animal odors. As a result of changes in traveling patterns, most guests arrived by steamboat, public stagecoach, and eventually railroad.[8]

Gaslight, introduced at an earlier date in Baltimore, was still enough of a novelty to lend an aura of modernity to its public rooms. A patented call-bell system enabled guests to signal the front office from their rooms, eliminating the annoying practice of bell ringing. And finally, the introduction of indoor plumbing accounted for the most sensational innovation. Reservoirs supplied the basement laundry, kitchen, and bathing rooms with piped cold water. Both guests and Boston residents alike patronized eight bathing rooms. The battery of eight indoor water closets located in a first floor passage provided an unprecedented introduction to high-tech luxury. Another thirty years would pass before anyone but the very wealthy could afford to acquire indoor plumbing. These innovations served to place the building in the forefront of early-nineteenth-century building technology. In an important step, Rogers integrated the technological systems into the design of the building, merging function with aesthetics while adapting historic architectural form and decorative detail to the practicalities of a modern, commercial building. This highly successful concept earned the hotel a national reputation from its first days. It initiated a standard by which all other houses were judged, and one that newer, more modern hotels sought to eclipse.[9]

The Tremont House retained its preeminence only until 1836, when John Jacob Astor opened the Astor House on Broadway in New York City. From that time forward New York City's hotels set the national standards for excess until Philadelphia's valiant effort in 1860 capped the antebellum years with the Hotel Continental. An article describing a new hotel in St. Louis in an 1863 issue of the London architectural journal *The Builder* chronicled the lineage of luxury hotels in the United States beginning with the Astor House. Each of the hotels mentioned "excelled," "eclipsed," "overtopped," "surpassed," and "cast in the shade" the best hotel of the immediate past, demonstrating a progressive competitiveness and the national diffusion of the idea of the luxury hotel as conceived originally by Rogers in his design of the Tremont House.[10]

Astor hired Isaiah Rogers to design his Astor House in the tradition of the Tremont House, but on a larger scale and with considerably more complex systems incorporated into the design. The New York hotel was a spectacular six stories, compared with the Tremont's four, contained nearly twice as many rooms, most larger than those in the Tremont, and cost twice as much money to build. The Astor had its own gas plant to supply gas for lighting, as well as a steam engine that provided power for the plumbing system, the laundry, and the kitchen equipment. Plumbing extended throughout the hotel, with hot and cold running water and water closets on all floors. The awesome power of steam drove one editor to note facetiously, "We believe it does not yet stipulate any assistance in bed-making, sweeping rooms, dusting furniture, attending on guests, &c., &c,; but in the onward march of improvement, we may expect all this to follow in good time." Other equipment included a printing press to print the daily menus, a bell system, and patent locks, all of which added to the hotel's reputation as "the greatest establishment of the kind in the world."[11]

Other hotels and inventions followed. The Metropolitan (1853) boasted of its boilers and steam engines, which occupied an "Engineer's Room." Steam heated all the public rooms and passages, radiating from pipes in the walls. The Metropolitan's laundry could process four thousand articles per day, supposedly having the ability to wash, dry, iron, and deliver a piece of laundry within the span of fifteen minutes. The St. Nicholas opened just a few months after, promising a laundry capable of turning out five thousand items per day. Gas and Croton water from the city's new water system coursed through the fifteen miles of piping contained in the hotel walls. New inventions included jacketed steam kettles and an "electro-magnetic" annunciator that replaced the older mechanical call bells. The *Daily Tribune* acknowledged the wild competitiveness of the industry when it predicted a "new era" for metropolitan hotels during which "they must be furnished without regard to cost."[12]

This spiraling development reached its pinnacle with the introduction of the passenger elevator by the Fifth Avenue Hotel in 1859. Paran Stevens, the hotel's manager, secured a design for a "vertical screw railway" from a Boston engineer, Otis Tufts. At the "station," off the main lobby, passengers entered an ornate "little parlor" with plush seats that carried them to the seven different floors above. While at first the elevator acted as an attraction in its own right, even something of a carnival ride, subject to erratic stops and starts, it eventually established itself as the feature that distinguished the more luxurious hotels from those of a lower stature. The elevator also had the impact of completely revamping the hierarchy of desirable rooms within the hotel. In the days of stair climbing, rooms on the lower floors were considered preferable and were most often reserved for women and families despite the annoyance of street noise and dust. Once the elevator alleviated the physical exertion necessary for reaching rooms on the higher floors, these rooms, formerly designated for men traveling alone, became the rooms of choice.[13]

During the thirty-five years between 1825 and 1860 the standards for size, opulence, and technological comfort in America's luxury hotels rose dramatically. Even so, the basic design set forth by Isaiah Rogers remained in place. The technology in hotels received detailed attention from newspapers and periodicals both in the United States and abroad. When advertising their hotels proprietors invariably described their amenities in terms of plumbing, steam heat, and annunciator systems. The technological profile of the hotel served a twofold purpose. First, it created a new definition of luxury, one embedded in the modern age. Because technology fit into the ideology of progress, it differed from its decorative counterpart. The interior decor, even while suffering from the vagaries of fashion, remained rooted in traditional old-world aristocratic styles. The new technological luxury, however, created its own destructive irony by continually turning against itself, creating obsolescence through improvements and forsaking what had been hailed as consummate achievements. Second, the technological image was symbolic. Cities exploited their hotels as symbols of commercial success and power within the national marketplace. Civic leaders grappling in increasingly competitive trade markets used the stunning hyperbole of the building as a means to attract visitors and commerce and to construct for their city a reputation as being uniquely civilized and refined. As the sociologist Judy Wajcman states, and it applies here in a very dramatic way, "To be in command of the very latest technology signifies being involved in directing the future."[14]

Having established a case for a technological and therefore powerful, masculine public image of the luxury hotel, it is time to move indoors and look at the way the design of the building accommodated men and women as they lived and worked. Geographers use a methodology based on visual perception that considers space another language for behavior. For example, to quote David Sack, "To explain why something occurs is to explain why it occurs where it does." David Harvey argues that space and social processes are two languages for discussing the same thing, in the same way that geometry and algebra use different math languages to express the same results. Feminist geographers explore the idea of gendered space in order to understand how, to use Daphne Spain's words, "initial status differences between women and men create certain types of gendered spaces and that institutionalized spatial segregation then reinforces prevailing male advantages." These ideas provide additional insights for understanding how the design of the luxury hotel translated the hierarchies between men and women and between classes into the built environment. The symbolic function of the hotel and its exaggerated proportions heightened the expression of social conventions and cultural values in its material form.[15]

An organizing principle of nineteenth-century American middle-class society was that of separate spheres, in which men and women assumed gender-specific roles stemming from the separation of work from the home and which evolved into dichotomies between private/home/women and public/work/male. While these oppositions are useful as a theoretical base for analysis, they can also obscure the many ways in which women entered the public domain and arenas where gender segregation was not quite so rigid. In this sense the luxury hotel both conforms to and departs from conventional expectations. The use of a domestic template to construct a public institution produced a building that incorporated both private and public spaces, such as bedrooms and parlors, except that at the hotel everything was greatly exaggerated. In addition, the wealthy culture of display that pervaded the hotel worked against middle-class prescriptions for modesty and against promiscuous gatherings and imparted a feminine character throughout the "house."[16]

The luxury hotel incorporated separate-sphere ideology into the design and use of space. Figure 2.1 shows the plan of the principal, or first, floor of the Tremont House. Guests passed under the front portico, entered through a pair of doors sixteen feet high, and ascended a ceremonious flight of ten steps to reach the stained-glass domed rotunda, at

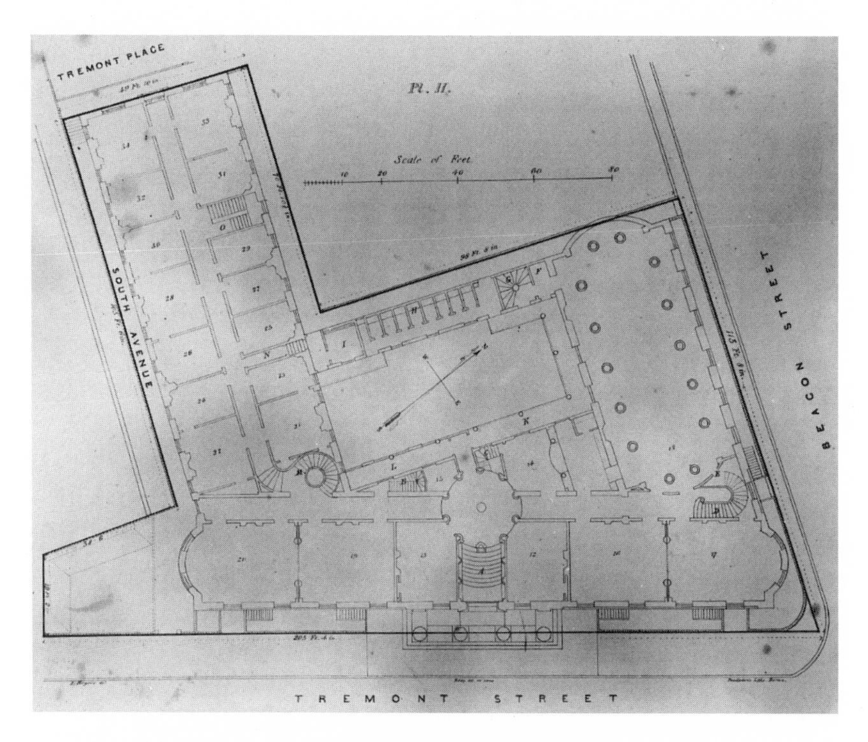

2.1. Plan of the principal floor, from *A Description of Tremont House* (Boston, 1830). Courtesy American Antiquarian Society.

which point women and men divided into their respective receiving rooms, women to the left (13), and men to the right (12). The space under the portico spanned the public world of the street and the private world of the hotel. Loose groups of "idle and dissipated" young men reputedly clustered under the portico, making it an uncomfortable place for respectable ladies to pass through unless accompanied by a male escort. Thus the experience of being swept up the grand staircase into the American commercial version of an aristocratic palace necessitated the sponsorship of a man, be it one's husband, a traveling companion, the doorman, or the hotel's proprietor. A separate entrance at the west end of the main corridor allowed women and men residents easy and more private access to the ladies' parlors and family suites.[17]

The first floor divided into men's and women's arenas. A series of doorways opened up from the rotunda. To the right along the front, the main corridor led to the men's drawing room, where the bar was located (16), and then the reading room (17). The reading room served as a sort of library, available by subscription to the (male) reading public, and held an assortment of national and foreign newspapers and periodicals. At the end of the corridor a staircase led to the upper floors of bedchambers in the north wing, rooms presumably filled by men traveling alone since women and families generally occupied the suites in the south wing. The main office (14) was situated across the corridor from the men's parlors, and the public dining room (18) occupied the first floor of the north wing. The dining room was the most spectacular room in the hotel. In Eliot's words, "As the largest and most public apartment of the house, it was considered deserving of the most elaborate decoration; and though the use to which it is devoted be not of a dignified or elevated cast, there seems at least no impropriety in surrounding its occupants with cheerful and tasteful objects." The hotel served four meals a day in the dining room: breakfast, dinner, tea, and supper; but only the men ate there, and so it anchored the male zone of the first floor, functioning with the other spaces as a venue for public ceremonies, the exchange of information, and the consolidation of social and business relationships.[18]

Returning to the rotunda, to the left the main corridor led to the women's spaces of the hotel. Directly to the left of the receiving room was the ladies' dining room (19), and to the left of that was the ladies' drawing room (20). A sliding door between them allowed women to pass without reentering the corridor and also, as in the other rooms, permitted a more flexible use of space. The south wing contained suites of private apartments in which two or three rooms could be connected as needed through the use of sliding doors (21–34). The same configuration was repeated on the second floor of the south wing, and the remaining floors were turned over to single chambers. The ladies' rooms equaled the men's rooms in size and were as prominently sited within the hotel, being located along the front of the first floor of the main building. However, it is important to note that these rooms were not used exclusively by women. Men dined with their families in the ladies' dining room, and women entertained their beaux, their friends, and relatives in the ladies' public parlors. Unlike the men's spaces, which excluded women and promoted a camaraderie of commercial, civic, masculine life, women's spaces more closely approximated those at home, where appropriately related men and women mingled and proper manners prevailed. Travelers occasionally commented upon the more pleasant atmosphere of the ladies' drawing rooms. One Englishman bemoaned that when traveling without a wife "you are shut out from many privileges, deprived of most agreeable society, and compelled to mourn your lone estate in company with fellows as wretched as yourself."[19]

The hotel also contained spaces in which neither sex benefited from guidelines for prescribed social behavior. These would include the corridors, the staircases, the piazza or covered walkway surrounding the inner courtyard, and the line of privies (H) situated in the corridor connecting the two wings behind the courtyard and parallel with the front main building. In an essay on women traveling alone during the early national period Patricia Cline Cohen observes that no preexisting etiquette existed to instruct men and women on codes of proper behavior. Since the Tremont offered the first public indoor water closets, one has to wonder what negotiations went on as men and women waited in line to use the facilities. Consider the problems women faced while using the facilities in "full dress." Or perhaps they chose not to use the water closets at all, preferring the privacy of chamber pots in their rooms. Published travel accounts almost never discuss personal hygiene except men's shaving experiences and the availability of bath water. As hotels grew larger and plumbing extended above the first floor, bathroom facilities became more private. By 1860 the new luxury hotels included private baths and water closets in the arrangements for family suites. Hotel barbershops eliminated the men's practice of shaving in the bar, where one could always count on there being a mirror. Also by midcentury etiquette books included special chapters on hotel behavior, institutionalizing a code of behavior that women had worked out through the experience of traveling alone.[20]

During the next thirty years hotels followed the basic plan of the Tremont House, but these buildings also grew tremendously in size, moving from the Tremont's 170 rooms to the Continental's 750 rooms. Variations were common. Nonetheless, in 1861 William Chambers reported that American hotels "consisted of two distinct departments—one for ladies and families, and the other for single gentlemen." Over this period of time the ladies' parlors became ever more luxurious. *Gleason's Pictorial* described the ladies' parlors of Philadelphia's Girard House, built in 1852, in gushing terms that fully conveyed the rooms' costly elegance. "Here expense has literally showered on the furniture and decorations. The floors are covered with painted velvet carpets, that echo no foot-fall; the curtains, yellow satin damask, relieved by rich lace hangings, and the most costly trimmings; sofas, lounges, etageres, tables &c., rosewood, *inlaid;* the sofas, &c., seated and backed with yellow satin, the chairs entire gilt, and yellow satin. The walls, from which gigantic mirrors blaze and multiply on every side, are decorated, and each parlor furnished with a massive chandelier of new style." When the Fifth Avenue Hotel opened in 1859 the ladies' parlors had been moved to the second floor, the first floor being given over entirely to men's space, from the telegraph office to the saloon to the grand entrance and exchange hall. At the Hotel Continental a similar arrangement provided an open balcony on the second story of the "grand stairway" that gave ladies "the opportunity of viewing the busy and enlivened scene below." The design encoded women's role as observers of men actively conducting the business of life in the hotel exchange. Moreover, it reified those aspects of separate-sphere ideology in which women observed from the sidelines the public activity of their men and served at the same time as moral watchdogs.[21]

As the size of luxury hotels increased, women's spaces grew proportionately and became ever more extravagant. Men, however, acquired new spaces, such as barbershops, billiard rooms, smoking rooms, telegraph offices, reading rooms, washrooms, and restaurants. Certain spatial changes occurred in concert with other cultural shifts. For example, the introduction of smoking rooms indicated a growing dissatisfaction with the intrusiveness of tobacco habits. Toward midcentury, dining practices also changed. The appearance of the street-level restaurant suggests both the development of a central business district that encouraged downtown eateries and a movement away from the traditional communal hotel table. Also about this time this time women gained regular access to the main dining room. A drawing in *Harper's Weekly* of the dining room of the Fifth Avenue shows large banquet tables seating twenty-five persons each. Men, women, and even a child in full evening dress sit at the tables and mingle in the aisles, crowding the room.[22]

The hotel's division of space clearly privileged men. Although women's spaces included men, men's spaces excluded women; thus, women were restricted to certain areas, while many men had the full run of the house. Because of the differentiation between men traveling alone and those traveling or visiting with women, however, there were actually three different gendered experiences within the hotel. Most significantly, married men, the true patriarchs, found all spaces within the hotel open to them. Their considerable cultural investment in both the economic and gender systems enabled them to symbolically and physically reign over the masculine public world and the feminized private domain. Male relatives and courters, patriarchs in training, also enjoyed these privileges. Thus the most powerful figure in American society, the patriarch, moved authoritatively through the one public building that enshrined both capitalist success and upper-class domesticity.[23]

Encircled by Corinthian columns and lit by enormous multi-globed chandeliers, the Fifth Avenue's dining room dripped with intricately ornate carvings, hangings, and mirrors, all in the latest French fashion. This was the way in which femininity asserted itself and pervaded the entire house: through the luxuriousness of the interior decoration. Consider the description of Phalon's Saloon, the barbershop at the St. Nicholas Hotel in New York City, printed in *Gleason's* in 1853. The furnishings of the "hair-cutting saloon" cost nearly $20,000. Presided over by Phalon, the "very *Barbiere di Seviglia* of New York," the saloon was lit by a "brilliant chandelier, . . . adorned on all sides by a splendid mass of mirrors," thirty in number ($8,000), frescoed ceilings ($3,000), and artistically carved rosewood chairs ($3,000). Each "artiste," or barber, wore an elegantly trimmed black velvet coat cut in the latest fashion. Magnificent washstands, statuary, marble floors, and a clean hairbrush for each lucky customer completed the ambiance. The article asserted that "the American who visits New York, and does not go to Phalon's Hair-cutting Saloon, is in infinite danger, during the next fifty years, of departing this life without having had the slightest idea of what it is to be shaved."[24]

Although this may seem to be an extreme example, in fact it is not. Most descriptions of hotels included some modest statement of preeminence similar to "It is furnished throughout with an elegance and sumptuousness unequaled by any hotel on this continent." The gentlemen's drawing room of the St. Nicholas was paved with marble and "decorated in the most elegant style." At the Continental the Gentlemen's Café boasted frescoed ceilings sixteen feet high. Thick imported carpets, enormous mirrors, sparkling chandeliers, satin and damask upholstery, rosewood furniture, velvet draperies, gilt frames, and countless other luxurious materials adorned the public rooms, the parlors and suites, and eventually the private bedrooms. Luxury in every form infused the hotel, lending it an intensely soft, feminine character that contrasted with the technological aspect of the building. The fostering of physical and sensory enjoyment and the concealment of productive activity contributed to the feminine character of hotel luxury. In contrast, the masculinity represented by the hotel's technology drew affirmation from material productivity and social, political, and economic accomplishment.[25] Thus, there was a profound tension between the hotel's feminized elegance and the building's ratification of masculine culture.

In what ways did men's behavior then reclaim feminized space as male? From the very start men used the hotel as a site for major political and civic dinners. The diary of Philip Hone, a wealthy and nationally prominent New York Whig, is a wonderful source for the kinds of events held in hotels. A dinner for Washington Irving, another for John Bell of Tennessee, several honoring Daniel Webster—these are just a few examples of gatherings of hundreds of men celebrating an event, a person, or a civic achievement. Smaller, personal dinners such as one Hone attended with Webster at the Astor House to experience the virtues of Merrimack River salmon, or another with "a select knot of four and twenty Whigs," also served to claim the hotel as a public home. In addition, politicians and popular foreign travelers often greeted and spoke to crowds from the front balconies of hotels. Not only did the balcony provide a heightened platform from which to speak but the hotel afforded a certain protection and the opportunity to retreat into private quarters when the speaker chose.[26]

There were ways in which men reacted against the effeminate luxury surrounding them. Habits such as smoking and chewing tobacco and, most offensive, spitting tobacco juice were notoriously prevalent during the first half of the nineteenth century. Several travelers commented on the incredible habit that many, though certainly not all, American men adopted of spitting tobacco juice on carpets and floors despite the numerous spittoons provided for their use. According to John Kasson, "Tobacco spit flew without restraint in places frequented by men: in railroad cars, stages, and steamers, saloons, hotels, and theaters, hospitals, colleges, and churches, shops, offices, private houses, post offices, courtrooms, state legislatures, the halls of Congress, and even the White House." Visitors decried the dense tobacco smoke that pervaded hotel dining rooms, drawing rooms, and barrooms. One letter writer suggested that smoking be banned from the Tremont House, submitting himself to public ridicule in the city's newspapers. Critics accused bachelors in particular of lounging about in complete disregard of the furniture, hoisting their feet on tables and laying down upon the sofas. Although most writers usually attributed this behavior to bad manners, and rightfully so, it may also have stemmed from a deliberate assertion of masculinity in the face of unrelenting social controls imposed upon them by the feminine character of their surroundings. Unable to "get away with it" in polite society, once in their own particular sphere some American men felt free, through a communal ownership, to indulge in such behavior. In this way they dominated space while at the same time exerting hegemony not just over the feminized environment but over women themselves, even in their absence.[27]

To summarize what I have said thus far, gender influenced the social construction of the hotel in three ways. First, the hotel's promoters defined the hotel in technological terms, creating a modern, progressive, and male identity for the building and for themselves. Second, the design of the hotel divided the structure into discrete zones according to sex. Different codes of behavior applied depending upon whether one was in the women's or the men's arenas. Third, the feminized luxuriousness of the decoration that permeated all guest areas of the hotel countered the technological identity and male dominance of space. The achievement of great wealth enabled rich men to indulge in leisure pursuits and surround themselves with refined and luxurious material goods.[28]

Gendered categories defined the hotel in yet another way: in the division within the institution between production (masculine) and consumption (feminine), or between the front of the house and the back of the house. The hotel is perhaps unique in this respect in that a veritable factory existed within the superstructure of the building to produce services that customers (literally and figuratively) consumed on site. Looking at the hotel in this way also introduces the element of class. An entire other class of people lived and worked in the hotel; this population was further divided between men and women, between free-born and immigrant, and by color. These workers straddled both worlds and provided the link between production and consumption. Very little is recorded about hotel workers in the antebellum years, yet considering them adds a new dimension to the exclusive atmosphere of the luxury hotel.

The production side of the hotel remained hidden from guests. At the Tremont the kitchen was located on the ground (basement) floor, underneath the dining room in the north wing (see fig. 2.1). A staircase (G) located between the privies (H) and the pantry (F) led from the kitchen to a hallway near the rear of the dining room. The servants' hall in the basement adjoined the kitchen and led to the office directly above by a staircase (C). Shops, accessible only from street entrances, occupied the ground floor space directly under the frontage of the building, on Tremont Street. The south-wing basement contained the laundry, the larder, and the eight bathing rooms. Men, usually Irish, served as waiters, while women worked in the laundry and housekeeping departments; thus the separation of these two work areas generally also served to separate workers by sex. The fourth and highest story of the main building held twenty-four rooms for boarding servants. Thus the building further defined the hierarchy of its inhabitants by placing its workers in the least desirable areas, the basement and the attic.[29]

Within less than a decade hotels introduced steam engines into their working departments, and so the production side became industrialized and more complex. The steam engine, introduced at the Astor as a "labor-saving" device, provided power to raise water above the ground level and made "itself generally useful" in the kitchen, the laundry, and the baths. A French traveler described a fourth wing, parallel to the front main building, where the servants lived, which also housed the mechanical departments such as the gas manufactory, the steam engine, the kitchen, and the laundry, which cleaned guests' clothing as well as house linens. This arrangement not only separated production from the hotel guests but also served a practical purpose by containing major fire hazards to one location. When the hotel first opened it employed eighty servants to accommodate guests staying in nearly four hundred rooms.[30]

When the Continental opened in 1860, it published a thirty-four-page booklet devoted largely to describing the mechanical plant of the building. The kitchen equipment included a twenty-two-foot long iron steam table used for carving and warming meats and gravies. Each waiter held responsibility for his own locked room full of silver, china, and glass. Steam heated nine boilers, and a separate little steam engine attached to the wall turned the roasting spits in the open fireplace. The kitchen used a gas-powered bread toaster. Steam power also ground coffee beans and froze ice cream. A separate refrigerator room cooled by ice kept meat fresh, while another cooled milk, butter, and fruit. A platform elevator transported stores from the receiving department below to the kitchen's work area. This listing represents just a fraction of the information supplied by the hotel describing the kitchen, the laundry, the plumbing and gas fittings, the heating system, and other "working" departments. The hotel had not only an exquisitely complicated mechanical plant but also an extensive labor force that included machinists, plumbers, and accountants in addition to the customary chambermaids and waiters. The Continental employed as many as three hundred workers, a number that the *Chamber's Journal* reported was necessary to take care of six hundred guests. In other words, the result of the new "labor-saving" devices was that a hotel required twice as many employees, one for every two guests, a trend that continued well into the twentieth century.[31]

In a nation that suffered perpetually from a servant problem the hotel often tried to sell itself as a solution. The continuous influx of new immigrants to the United States kept hotels supplied with labor. Travelers noted that in the North waiters were usually "Irish lads," the chambermaids and laundry workers, young Irishwomen. A trend developed in the North to hire free blacks as waiters and runners, and by 1860 that custom seemed entrenched in the very best hotels. In the South hotels used slaves, sometimes owned by the proprietors but most usually hired. Even so, the very best hotels in the South eventually began to employ Irish immigrants rather than slaves. The chefs were usually French, or at least European. As openings for clerks, bookkeepers, and other lower-management positions developed at the hotels, these were filled by native-born Americans, mostly young New Englanders training for a hotel profession. The positions of barkeeper and proprietor enjoyed a certain respectability, much to the ongoing surprise of foreign visitors. Travelers often noted that native-born Americans refused to accept service positions because it ran counter to democratic ideology, "their pride not allowing them to fill such places," as one Englishman put it. By 1860 the hotel world really was a microcosm of American society, not only gendered but filled with people from the very wealthy to the newly arrived immigrant, blacks and whites, craftsmen and artisans, skilled and unskilled workers, merchants and storekeepers, and the emerging class of middle management, each with their appropriate door for going in and going out.[32]

At the most basic level of consumption travelers and permanent residents engaged rooms in exchange for payment of the going rate, which at this period of time ranged from two dollars per day for a single room to about six dollars per day for a suite, with more reasonable weekly rates available. This cost included four meals per day; wine and other liquors, laundry, baths, stationery, fires, and other services were extra. However, people who stayed at urban luxury hotels did so because they received more than just a place to sleep and eat. The hotel as an American version of a European aristocratic palace reinforced or upgraded one's station in life. As the historian Dona Brown stated, "Industrial capitalism has been able to encompass the buying and selling of cultural experiences that seem to be outside it, or even in direct conflict with it." In this case the hotel emerged as a place for the commodification of an upper-class identity, a kind of "rent-a-palace."[33]

When one stayed at a luxury hotel, one was purchasing comfort. Not only did one live amidst great splendor but overwhelming food choices awaited in the dining room and the hotel provided a level of physical well-being most likely unavailable at home. From the early water closets and bathing facilities to steam heat, private bathroom facilities, and gas lighting, the availability of technological luxury made staying or living at the hotel an exceptional experience. In this regard technology as an experience was equally available to both men and women. In fact women may have benefited more from technological luxury. The elevator enabled women to reserve rooms on the upper floors, thus sparing them the noise and dirt associated with the rooms on the first and second floor to which stair climbing had relegated them. Family suites were the first to acquire private bathroom facilities. The greater demands for privacy that accrued to women also resulted in greater convenience.

As a self-contained city the hotel cocooned its guests in an atmosphere divorced from the chaos of the world outside. Astor enlarged upon the concept of renting retail space to merchants. The stores at the Astor House communicated both to the street and the interior of the hotel. Thus, travelers had access not only to retail goods such as clothes, shoes, apothecary items, and umbrellas but also to services such as the post office and a ticket office. In describing the Astor the "Frenchman" noted that the hotel was "a city in the midst of a great city," a metaphor that continues to be used to describe large hotel complexes today.[34]

Despite its protective insularity, the hotel also served as a gateway to the city for women who enjoyed the convention of display on Broadway, shopping, and the theater. However, just as the spatial dimensions of the hotel controlled women's movements, these diversions too fell within carefully drawn boundaries that defined acceptable urban activities for women. Many of the New York City hotels arrayed themselves along the stretch of Broadway known as the Ladies Mile, from Fourteenth Street to Twenty-third. The Fifth Avenue, at the intersection of Broadway, Fifth, and Twenty-third Street at the south end of Madison Square, anchored one end. Stores such as the fabulous A. T. Stewart, at Broadway and Chambers Streets, reproduced the luxurious structure of the hotel in the form of a shopping emporium, complete with mirrored and gilded ladies' parlors. Commerce welcomed women to the city, leading them from the hotel to the department store. A city's finest luxury hotel situated itself in the center of a city's commercial district, permitting easy access for both men and women. Women found this particularly important because they did not share the same unrestricted freedom that men had to roam the city.[35]

Beginning with the Astor House and continuing beyond the period of this chapter, architects designed the corridors along the drawing rooms and parlors as wide avenues with seats and mirrors and shimmering lights, replicating in a controlled and luxurious fashion the city streets outside. William Chambers called them "flirtation-galleries on account of their qualities as places of general resort and conversation." The Frenchman was less coy in his description: "*On dit* the great hotels in America produce alone more marriages than any private society. These establishments, therefore, have a great importance in a country where the increase of the population is considered as one of the principal elements of its prosperity." In an editorial deploring the privatization of dining being experimented with at a new uptown hotel the *New York Weekly Mirror* protested, "The going to the Astor, and dining with two hundred people, and sitting in full dress in a splendid drawing-room with plenty of company—is the charm of going to the city! The theatres are nothing to that! Broadway, the shopping and the sights, are all subordinate—poor accessories to the main object of the visit." Reflected in the wall-to-wall glittering mirrors, young women displayed themselves as magnificently as the consumer goods on Broadway were displayed. Indeed their costumes served to attract attention to them just as surely as the parlors framed them in a theater set designed to flatter. Ultimately, one consumer product manufactured by the hotel's network of capitalist enterprise was the young capitalist couple to be, the heirs of commercial success and the consolidators of future empires.[36]

Commercial luxury hotels were important institutions in nineteenth-century American cities and commanded a far more conspicuous and symbolic role than do our own late-twentieth-century hotels. Their emergence coincided with other great historical developments, such as industrialization, technological change, the broadening of political rights, the solidification of separate spheres, burgeoning consumerism, and the gentrification of the middle class. Luxury hotels were both a product and a source of these profound changes. In their size, technological development, interior decoration, and service these buildings magnified urban life and encouraged the interplay between its customers, the hotel, and the city. While they had detractors (*Harper's Weekly* in particular led a campaign against young married couples setting up housekeeping in hotel), civic leaders everywhere enthusiastically embraced the idea of the luxury hotel as a leading indicator of cosmopolitanism and commercial success.

Technology, gender, and consumption all contributed to the social construction of the hotel. Innovations defined a modern genus of luxury, what I call technological luxury, that not only provided new experiences of comfort but contributed to the hotel's representation as a progressive urban artifact. The traditional, feminine luxury inherent in the interior decoration contributed to the hotel's definition as a palace and helped to make the hotel a hospitable place for women even as it subtly invoked the tensions of nineteenth-century manhood. Both forms of the hotel's luxury expanded the parameters of consumption, enabling guests to both purchase experience and enlarge their own visions of "need" and consumptive power. Luxury, then, became the nexus connecting technology, gender, and consumption. That the luxury hotel served as a symbol for democratic America signified the power of economic forces within an emerging industrial nation.

Notes

Research for this chapter was supported by grants from the American Antiquarian Society, in Worcester, Massachusetts, and the National Museum of American History, in Washington DC. I also wish to thank Angela Woollacott, Catherine Kelly, Carroll Pursell, and the Works-in-Progress Group at Case Western Reserve University for their perceptive comments.

1. Abraham Lincoln, "The House Divided Speech," in *The American Reader: Words That Moved a Nation,* ed. Diane Ravitch (New York: HarperCollins, 1990), 118–19. Lincoln in particular is known for using the Willard Hotel in Washington DC as a headquarters (see Ledger for 1861, Bill Book No. 3, Willard Family Papers, Library of Congress).

2. Michael Dennis, *Court and Garden: From the French Hotel to the City of Modern Architecture* (Cambridge: MIT Press, 1986), 3; Richard L. Bushman, *The Refinement of America: Persons, Houses, Cities* (New York: Alfred A. Knopf, 1992), 61–64; Jane Boyle Knowles, "Luxury Hotels in American Cities, 1810–1860" (Ph.D. diss., University of Pennsylvania, 1972), 3 n. 3; Doris Elizabeth King, "Hotels of the Old South, 1793–1860" (Ph.D. diss., Duke University, 1952), 1–10; Meryle Evans, "Knickerbocker Hotels and Restaurants, 1800–1850," *New York Historical Society Quarterly* 36 (Oct. 1952): 383; "Palace Homes for the Traveller," *Godey's Lady's Book and Magazine* 60 (May 1860): 465–66. For a contemporary usage of the phrase see J. M. Fenster, "Palaces of the People," *American Heritage* 45 (Apr. 1994): 46–58.

3. Paul Johnson, *The Birth of the Modern: World Society 1815–1830* (New York:

Harper Collins, 1991), xvii; Jefferson Williamson, *The American Hotel: An Anecdotal History* (New York: Alfred A. Knopf, 1930), 13; Kermit L. Hall, *The Magic Mirror: Law in American History* (New York: Oxford Univ. Press, 1989), 97–98.

4. Gillian Rose, *Feminism and Geography: The Limits of Geographical Knowledge* (Minneapolis: Univ. of Minnesota Press, 1993), 35; Leslie Kanes Weisman, *Discrimination by Design: A Feminist Critique of the Man-Made Environment* (Urbana: Univ. of Illinois Press, 1993), 86; Bushman, Refinement of America, xvii; Philip Hone, for example, often referred to the times as the "go-ahead age" (see Philip Hone, *The Diary of Philip Hone, 1828–1851*, ed. Allan Nevins, 2 vols. [New York: Dodd, Mead, 1927], 261, 410, 714).

5. Thomas Hamilton, *Men and Manners in America*, 2 vols. (Philadelphia: Carey, Lea & Blanchard, 1833), 1:90; Peter Dobkin Hall, *The Organization of American Culture, 1700–1900: Private Institutions, Elites, and the Origins of American Nationality* (New York: New York Univ. Press, 1984), 19, 100; Frederic Cople Jaher, The Urban Establishment: Upper Strata in Boston, New York, Charleston, Chicago, and Los Angeles (Urbana: Univ. of Illinois Press, 1982), 25, 50.

6. Weisman, *Discrimination by Design*, 9; list of investors found in Benjamin F. Stevens, "The Tremont House: The Exit of an Old Landmark," *Bostonian* 1 (Jan. 1895): 331 (nearly all the investors lived within less than a half-mile radius of the hotel site); Abel Bowen, *Bowen's Picture of Boston*, 2nd ed. (Boston: Lilly Wait & Lorenzo Bowen, 1833); William Havard Eliot, *A Description of Tremont House* (Boston: Gray & Bowen, 1830), 5.

7. On the social construction of technology as masculine see Carroll Pursell, "The Construction of Masculinity and Technology," *Polhem* 11 (1993): 206–19; Arnold Pacey, *The Culture of Technology* (Cambridge: MIT Press, 1983), 100–101; Judy Wajcman, *Feminism Confronts Technology* (University Park: Pennsylvania State Univ. Press, 1991), 17, 22, 49; Michael Adas, *Machines as the Measure of Men* (Ithaca: Cornell Univ. Press, 1989), 14; Gillian Brown, *Domestic Individualism: Imagining Self in Nineteenth-Century America* (Berkeley: Univ. of California Press, 1990), 23; For machinelike descriptions see Tyrone Power, *Impressions of America; During the Years 1833, 1834, and 1835*, 2 vols. (Philadelphia: Carey, Lea & Blanchard, 1836), 1:78; William Chambers, *Things as They Are in America* (1854; reprint, New York: Negro Universities Press, 1986), 185.

8. Bowen, *Picture of Boston*, 286–88. That the Granite was the first railway is disputed by Frederick C. Gamst, "The Context and Significance of America's First Railroad on Boston's Beacon Hill," *Technology and Culture* 33 (Jan. 1992): 66; see also Williamson, *The American Hotel*, 22, and Bowen, *Picture of Boston*, 218.

9. Eliot, *Description of Tremont House*, 13–16, 35–36; Samuel Eliot Morison, *One Boy's Boston, 1887–1901* (Cambridge MA: Riverside Press, 1962), 4; Susan Strasser, *Never Done: A History of American Housework* (New York: Pantheon, 1982), 96–97.

10. Thomas Walsh, "'The Lindel,' of St. Louis," *The Builder* 21 (7 Feb. 1863): 92–93.

11. Williamson, *The American Hotel,* 265; Kenneth Wiggins Porter, *John Jacob Astor: Business Man,* 2 vols. (Cambridge: Harvard Univ. Press, 1931), 2:994–95; Vaughn L. Glasgow, "The Hotels of New York City prior to the American Civil War" (Ph.D. diss., Pennsylvania State University, 1970), 70–77; "Astor House," *New Yorker,* 4 June 1836, 173; "The Large Hotel Question," Chamber's Journal 21 (1854): 153.

12. *New York Herald,* 2 Sept. 1852, as cited in Glasgow, "Hotels of New York City," 91; "Large Hotel Question," 153; "St. Nicholas Hotel, New York," *Gleason's Pictorial,* 12 Mar. 1853, 161; "The St. Nicholas Hotel," *New York Times,* 7 Jan. 1853; *New York Daily Tribune,* 3 Dec. 1852.

13. Cecil D. Elliott, *Technics and Architecture: The Development of Materials and Systems for Buildings* (Cambridge: MIT Press, 1992), 330–31; "Steam versus Stairs," *New York Times,* 23 Jan. 1860.

14. Wajcman, *Feminism Confronts Technology,* 144.

15. Robert David Sack, *Conceptions of Space in Social Thought* (Minneapolis: Univ. of Minnesota Press, 1980), 70, as quoted in Daphne Spain, *Gendered Spaces* (Chapel Hill: Univ. of North Carolina Press, 1992), 5; David Harvey, *Social Justice and the City* (Baltimore: Johns Hopkins Univ. Press, 1973), 46–47; Spain, Gendered Spaces, 6. See also Weisman, *Discrimination by Design*; Rose, *Feminism and Geography;* and Sara Deutsch, "Reconceiving the City: Women, Space, and Power in Boston, 1870–1910," *Gender and History* 6 (Aug. 1994): 202–23.

16. Carroll Smith-Rosenberg, *Disorderly Conduct: Visions of Gender in Victorian America* (New York: Oxford Univ. Press, 1985), 60, 88–89; Linda Kerber, "Separate Spheres, Female Worlds, Woman's Place: The Rhetoric of Women's History," *Journal of American History* 75 (June 1988): 13; Deutsch, "Reconceiving the City," 208.

17. Eliot, *Description of Tremont House,* plate 2; Boston Commercial Gazette, 19 Oct. 1829.

18. Eliot, Description of Tremont House, plate 2, pp. 11–12; Power, *Impressions of America,* 1:77–80.

19. Carolyn Brucken, "The Gender Organization of Hotel Space" (paper presented at the annual meeting of the American Studies Association, Pittsburgh PA, 10 Nov. 1995), 3; Charles Augustus Murray, *Travels in North America During the Years 1834, 1835, and 1836,* 2 vols. (London, 1839), 1:62, 102; "The St. Nicholas Hotel, New York," *Graham's Illustrated Magazine,* Feb. 1857, 167–68.

20. Patricia Cline Cohen, "Safety and Danger: Women on American Public Transport, 1750–1850," in *Gendered Domains: Rethinking Public and Private in Women's History,* ed. Dorothy O. Helly and Susan M. Reverby (Ithaca: Cornell Univ. Press, 1992), 110; James Stuart, *Three Years in North America,* 2 vols. (New York: J. and J. Harper, 1833), 1:87. For etiquette books see, e.g., Florence Hartley, *The Ladies' Book of Etiquette, and Manual of Politeness, A Complete Hand Book for the Use of the*

Lady in Polite Society (Philadelphia: G. G. Evans, 1860), 40–43.

21. Chambers, *Things as They Are in America,* 179; " House, Philadelphia," *Gleason's Pictorial,* 21 Feb. 1852, 114; "The Fifth Avenue Hotel," *New York Times,* 23 Aug. 1859; *Traveler's Sketch* (Philadelphia: Continental Hotel Company, 1861), 13, Hagley Museum and Library, Wilmington DE; Nancy F. Cott, *The Bonds of Womanhood: "Woman's Sphere" in New England, 1780–1835* (New Haven: Yale Univ. Press, 1977), 148–49. I am indebted to Catherine Kelly for this insight.

22. "Phalon's Saloon," *Gleason's Pictorial,* 5 Feb. 1853, 112; "Fifth Avenue Hotel"; *Harper's Weekly,* 1 Oct. 1859; "The St. Nicholas Hotel, New York."

23. I am indebted to Angela Woollacott for these insights.

24. M. Christine Boyer, *Manhattan Manners, Architecture and Style, 1850–1900* (New York: Rizzoli, 1985), 53; "Phalon's Saloon," 112.

25. *Guide to Philadelphia* (Philadelphia: John Dainty, 1868), 125–26; "The St. Nicholas Hotel"; "Traveler's Sketch," 11. On the relationship between luxury and effeminacy see Christopher J. Berry, *The Idea of Luxury: A Conceptual and Historical Investigation* (Cambridge: Cambridge Univ. Press, 1994).

26. Hone, *Diary,* 2:628, 703. See also *Gleason's Drawing-Room Companion,* 27 Dec. 1851, 553; 11 Oct. 1851, 373; 17 May 1851, 37; 7 June 1851, 86.

27. Henry Tudor, *Narrative of a Tour in North America* (London, 1834), 421–22; Frances Trollope, *Domestic Manners of the Americans* (1839; reprint, London: Century, 1984), 12; John Kasson, *Rudeness and Civility: Manners in Nineteenth-Century Urban America* (New York: Hill & Wang, 1990), 125; E. T. Coke, *A Subaltern's Furlough,* 2 vols. (New York: J. and J. Harper, 1833), 1:34; *Boston Commercial Gazette,* 19 Oct. 1829; *Harper's Weekly,* 26 Dec. 1857, 825.

28. T. J. Jackson Lears describes a consequence of luxurious living, the loss of masculinity with which upper-class men were associated by virtue of their separation from "real" productive work. The benefits of leisure also threatened to "enfeeble" those who submitted to an intemperate lifestyle (see T. J. Jackson Lears, *No Place of Grace: Antimodernism and the Transformation of American Culture, 1880–1920* [1983; reprint, Chicago: Univ. of Chicago Press, 1994], 56–57).

29. Knowledge of the composition of the work force is gained from the published traveler's accounts of foreign visitors, who, used to an established servant class, found the system, or perhaps the lack of a system, bemusing if not incomprehensible (see esp. Eliot, *Description of Tremont House,* plate; E. T. Coke, *A Subaltern's Furlough,* 1:32; Power, *Impressions of America,* 1:80; Trollope, *Domestic Manners of the Americans,* 489.

30. "Astor House," 173.

31. "Traveler's Sketch," 15–23; "The Large Hotel Question," 153; "Mechanical Equipment Boosts the Payroll," *Hotel Monthly* 31 (Nov. 1923): 21.

32. Power, *Impressions of America,* 1:79; Trollope, *Domestic Manners of the Americans,* 489; G. A. Sala, "American Hotels and American Food," *Temple Bar Magazine* 2 (July 1861): 355; "Metropolitan Hotel, New York," *Gleason's Drawing-Room Companion,* 25 Sept. 1852, 201; King, "Hotels of the Old South," 266–74; Coke, *A Subaltern's Furlough,* 1:32.

33. Hotel bill of William Bayliss, 25 June 1831, Misc. File, Massachusetts Historical Society, Boston; bill of J. A. Rockwell, 1842, Lithography File, American Antiquarian Society, Worcester MA; Dona Brown, *Inventing New England: Regional Tourism in the Nineteenth Century* (Washington DC: Smithsonian Press, 1995), 13.

34. "A Frenchman's Idea of the Astor House," *New Mirror* 1 20 (19 Aug. 1843): 311; "Astor House," *American Traveller,* 14 June 1836. For a contemporary use of the "city within a city" metaphor see, e.g., "New York City to Rise in Las Vegas," *New York Times,* 7 May 1995.

35. Boyer, *Manhattan Manners,* 43; Mary Ryan, *Women in Public: Between Banners and Ballots, 1825–1880* (Baltimore: Johns Hopkins Univ. Press, 1990), 76.

36. Chambers, *Things as They Are,* 182; "A Frenchman's Idea of the Astor House," 311; "New Kind of Hotel Up Town," *New York Weekly Mirror,* 7 Dec. 1844.

Three

Love, War, and Chocolate
Gender and the American Candy Industry, 1880–1930

Gail Cooper

An enthusiasm for machinery pervaded American industrialization. So it is not surprising that when mechanization and mass production came to the American confectionery industry at the turn of the century one industry journal boasted, "No more pointed or striking illustration of modern adaptation of machinery could be asked or given than the plain statement that nearly all the candy consumed to-day throughout the United States is made by machinery."[1]

We are inclined to think of Henry Ford's Model T as the quintessential example of American mass production, by which machine manufacture produced consumer goods in enormous quantities and brought their cost within the reach of the average person.[2] However, Ford was a sometimes unsatisfying apostle of modern consumer culture. He did not entirely approve of stylistic changes, advertising, or consumer credit, and he promoted saving rather than spending among his employees. In contrast, Alfred Sloan, of General Motors, knew precisely how to sell style and status to the American car buyer and presided over the invention of the annual model change, a form of planned obsolescence. Sloan understood that mass production was necessarily linked to mass consumption. And how those links were made—through advertising, brand-name products, and the creation of novelty—has drawn increasing scrutiny from historians.[3]

But how would our understanding of industrialization change if the chocolate bonbon, instead of the automobile, became our symbol of mass production? Instead of Diego Rivera's brawny masculine auto workers transforming the raw steel of Ford's River Rouge plant into a sleek roadster with BBD&O's carefree Dad slipping behind the wheel, we could illustrate our histories with slender young women in spotless white tending a seemingly endless belt of glistening candies with an eager housewife or working girl waiting to purchase her favorite confection.[4] In short, we would see women instead of men as both the workers and the consumers in the story of mass production.

That explicit recognition of gender opens the door to a more complex understanding of both mass production and mass consumption. Acknowledging the power of gender roles to shape industrial development and business strategies is a natural outgrowth of our understanding of the myriad ways in which it shaped personal experience. Indeed, in the years from 1880 to 1930 manufacturers and workers alike understood that there was no such thing as a gender-neutral "factory operative." Instead there were male workers and female workers, for theirs was a world of men's jobs and women's jobs, of men's wages and women's wages. In an analogous way consumer patterns were gendered as well. Women's economic dependence on men, both prescriptive and real, was a defining characteristic of their consumer identity. Women never escaped the consequences of being female, and their collective personal stories make up the history of the candy trade.

In many respects mechanization in the confectionery industry could stand as a perfect example of the symbiotic relationship between mass production and mass consumption. As manufacturers transformed their factories, by necessity they recast their relationship to the consumer as well. Consumer loyalty to packaged, brand-name products allowed a higher profit margin to pay for the expense of mechanization, and the acceptance of candy as an everyday indulgence increased levels of consumption to match the steady output of factory production. Yet this understanding of the links between production and consumption would be incomplete if we failed to realize that these manufacturers lived in a gendered world. Confectionery was called a "woman industry" because a majority of its workers and, increasingly, a number of its consumers were female. These workers and consumers viewed themselves as connected not only by economic exchanges but by sisterhood as well. Disenfranchised until after World War I, women saw in consumer power a potent tool for social change. Exploiting the importance of packaged, brand-name products to the new consumer-producer nexus, the National Consumers' League rallied women buyers to employ the threat of boycott to effect industrial reform for female factory workers.

Inventing a Mass Production Industry

In 1880, before mechanization swept the industry, most candy was made by hand by skilled male workers. Sugar was the main ingredient, and variety was created by the way in which it was heated and worked; technique rather than ingredients was the key to variety. The amount of heat necessary to achieve the right consistency required judgment, and the physical manipulation of the batch after it was taken out of the kettle called for strength and skill. Candies that needed to be worked after cooking were often put on heavy hooks and "pulled." As a twenty-pound mass of candy stretched from its own weight a distance of six to eight feet toward the floor, it was "gathered up quickly and accurately by a deft worker, who would, by a dexterous movement, give the two sections an upward toss that would land the flexible sweetness squarely on the hook again and continue the 'pulling.'"[5] Experts estimated that it took a man seventeen minutes to pull twenty pounds of candy.[6]

Hand methods first gave way to mechanization between 1880 and 1900. "The idea, of course," explained one industry journal, "is to do the work automatically. One machine with an expert and two assistants is said to do the work of 65 ordinary workers."[7] Mechanization came with large-scale enterprise and a modernist perspective, captured by the comment of one industry writer who declared, "The homelike expedients of the small makers, of the retail dealers who manufacture on their premises, and of the tiny shops where one can buy what is avowedly 'old-fashioned home-made candies' have no place in the economic plant of the large manufacturer. Here machinery and method rule the day."[8] Production now took place in factories, which were distinguished by approach, scale, and product from older workshops. One commentator boasted that "the most modern and expensive machinery has been added to the equipment of up-to-date factories," and when one such moderate-sized factory was put up for sale in 1905 the owner estimated that he had spent more than seventy-six thousand dollars to build and equip it.[9]

Despite this expensive investment in machinery, the price of candy to the consumer declined. Indeed, one journal claimed that prices fell because of mechanization.[10] In the words of another writer, "Improved machinery, the result of American inventive ingenuity, has so lessened the cost of production that it is now possible to purchase pure, wholesome candies as low as 10 cents a pound at retail."[11] Ten cents was a dramatic contrast to prices in 1876, when "American mixed" candy sold for thirty cents a pound and plain French creams sold for sixty cents. Indeed, there was a phenomenal rise in candy consumption in this period: annual per capita consumption of candy in the United States rose from 2.2 pounds in 1880 to 5.6 pounds in 1914, an increase of 250 percent in thirty-four years.[12]

Not only did consumers benefit from lower prices but the varieties of candy exploded. One writer remembered that in 1885 "the molasses candy, stick candy, sour balls, gum drops and peppermints constituted the stock-in-trade of the stores." Two decades later more elegant creations competed for consumer dollars. "The plebeian molasses candy still sells," he conceded, "but it is shouldered out of place by such aristocrats as chocolate creams, bon-bons and marrons glaces." Not only were candies more elaborate but the sheer variety was staggering; the New England Confectioner Company, of Boston, made more than five hundred different varieties of candy in 1911.[13]

The story of candy as told by contemporaries was a classic tale of mass production. Mechanization reduced labor, increased output, lowered costs, encouraged variety, supported economies of scale, and created a modern factory operation in which "machinery and method" ruled. In these aspects the candy industry is not surprising. But a closer look reveals that mechanization was not independent of changes in consumption.

Despite the vast output of modern American factories, profits in the confectionery industry were not always certain. Often manufacturers tried to secure customers by competing on the basis of price.[14] Penny candy, however, left little room for the manufacturer to squeeze out profits. The problem was particularly acute among southern manufacturers, sellers to a rural clientele who favored the cheapest varieties. One engineer trying to find a more efficient production system for a small firm described their dilemma: "The Southern candy factories are all burdened with a big stick candy business, which, on the whole, is not considered very profitable. . . . The fact that it is a staple in every line makes the price so low that many times if you are 'swapping' dollars you are doing very well on this product."[15]

Selling at cost left few profits for modernization, and machinery makers knew it. Perhaps the most vocal and the most successful of the American machinery makers was Frank L. Page, president of the Confectioners' Machinery and Manufacturing Company. Established around 1893, the company employed several hundred workmen in a foundry, machine shop, and woodworking shop at a four-acre site in Springfield, Massachusetts. By 1909 the company was manufacturing more than fifty different machines, each selling for between one hundred dollars and four thousand dollars.[16]

In 1905 Page addressed the problems of manufacturers. He condemned those who perpetuated both old-fashioned production methods and an outmoded view of business relations. "Too many of you follow ancient methods and refuse to be converted," he scolded. "If you do discover some new method of cheapening your product, you at once want to cut the price, instead of profiting by your discovery." He advised candy manufacturers to ensure their own prosperity through a variety of merchandising efforts. "Use your brains a little more," he exhorted them, "find 'something different' in the way of new novelties, new packages, new selling arrangements, new appliances. . . . In short, look for 'something different,' so you won't have to feel that it is necessary to cut prices to get business." It is clear that Page saw an explicit connection between new methods of production and new patterns of consumption; only innovative merchandising would pay for the expensive machinery he wanted to sell manufacturers.[17]

Page was articulating one of the important functions of advertising. Modern advertising methods were instrumental not simply in selling the greater volume of goods pouring out of newly mechanized factories but also in breaking the connection between cost and price. In Page's view price competition essentially was destructive. His concern with the distress of manufacturers caught in price competition suggests that mechanization did not automatically ensure prosperity for the businessman through reduced labor costs. Instead it was necessarily linked to new methods of selling.

It soon became clear that new patterns of consumption were crucial to survival of firms that had already mechanized. The confectionery industry struggled with a traditional pattern of seasonal production that was a poor fit with heavy year-round capitalization costs. The candy industry rode a cycle of intense activity in the weeks before Christmas followed by a steady decline until summer brought production nearly to a halt. This seasonal pattern was based on two factors, the hot humid summer weather that made production of weather-sensitive candies nearly impossible and the peak consumer demand tied to the holiday season. Seasonality, a reality of nature that found numerous expressions in American culture, was well entrenched in the social calendar and was the bane of the new production systems, which were relentlessly constant.[18]

Manufacturers first applied technology to the problem of summer shutdowns. Air-conditioning systems created an artificial environment within the factories that made them independent of seasonal weather and its problems. New packaging kept confections insulated until they reached the consumer. But manufacturers needed to stimulate year-round demand to match their year-round capacity, and modifying consumers' buying patterns presented a more complex problem than adopting simple technical solutions.

The industry's first effort can best be described by what anthropologist Sidney Mintz calls "intensification." As society adopts a new food, Mintz argues, its consumption acquires meaning through either "intensification," an emulation of older patterns, or "extensification," a wider usage that gives it a new character. Industry leaders first tried the former. Building upon the familiar Christmas pattern, the industry journal *Confectioners' Gazette* explicitly promoted greater consumption by reinforcing the link between candy and holiday celebrations; it simply urged the industry to extend the practice to a greater number of holidays.[19]

By 1914 the industry's merchandising efforts succeeded in establishing

Easter as a second major holiday associated with novelty candy. One industry journal commented on the new Easter peak: "It is a great time of rejoicing in the church year, and people show their elation in the purchase of large quantities of confectionery. . . . This busy time among confectioners is comparatively of recent growth."[20] Easter candy followed an old tradition of molding sugar into ornamental and representational forms that the modern industry called "novelties." Easter eggs were a European tradition, and Philadelphia was a notable center for their production by the 1880s. There one could find crystal-shell eggs, for instance, made by filling an inverted egg mold with hot syrup and decorating the cooled egg with lithographed cutouts. Such tediously crafted and costly confections were increasingly replaced by cream eggs, made by casting the shape in starch, a mechanized process. One proponent of merchandising pointed out the display value of Easter novelties: "As is the case in displaying special candies at other times, so is it prior to Easter; that is, a number of special candies such as will appeal for trade on account of their seasonableness."[21]

Industry strategy centered on heightening the consumer's sense of seasonality and providing a succession of such holidays. The *Confectioners' Gazette* remarked, "St. Valentine's Day and Washington's and Lincoln's birthdays also give the progressive confectioner many opportunities for materially helping trade along in the right direction."[22] Thus the industry tried to capitalize upon the American social calendar by stressing candy's celebratory character. In 1916 the National Confectioners' Association declared an annual Candy Day to create one more occasion to extol and consume confections. "Candy is closely interwoven with the world's onward march of civilization," declared one supporter. "Imagine, if you can, a courtship carried on without confections, a birthday without a box of bonbons, a Christmas without candy."[23]

Candy's social significance as a marker of the special occasion was clearly seen in its importance to courtship. Men routinely bought candy for the women they courted. Mintz marvels that "one male observer after another displays the curious expectation that women will like sweet things more than men; that they [men] will employ sweet foods to achieve otherwise unattainable objectives." The industry was well aware of men's importance as consumers and the heavy burden that candy carried in expressing a suitor's serious intent, stylish sensibilities, and social status. To proponents of merchandising, courting men were the perfect consumers for they valued presentation before price. Suitors formed a ready market for the new packaged products that gradually supplanted bulk goods.[24]

Ultimately the multiplication of holidays and special-occasion candies was not the most successful strategy for expanding markets. The largest increase in candy consumption resulted from its transformation from a special-occasion treat to an everyday indulgence. In Mintz's terms, candy followed a process of extensification rather than intensification. It was not love but war that expanded consumption among American men. World War I offered confectionery manufacturers an army of captive consumers and a chance to shape their eating habits.

During World War I candy became part of military rations for the first time, supplanting simple molasses, sugar, or syrups. The recognition of candy's value as a high-caloric, quick-energy food began during the Boer War. The American confectionery industry watched with great interest as candy was transformed from a luxury indulgence to a dietary staple. In 1903 the *Confectioners' and Bakers' Gazette* reported: "An invigorating effect is realized from administering in any form a very liberal amount of sugar. This fact was more than demonstrated during the late war in South Africa when one of the most difficult of modern campaigns was fought by the British on rations consisting largely of jams, chocolates, and other sweets." Indeed, playwright George Bernard Shaw also took note of this development: in *Arms and the Man* the mercenary soldier Bluntschli confides, "I've no ammunition. What use are cartridges in battle? I always carry chocolate instead."[25]

When the U.S. Quartermaster General's office began centralized purchasing of candies through the Food Administration in 1917 they favored a limited range of hard candies: lemon drops and mixed candies, which were to contain 50 percent peppermints. Although they constituted only 15 percent of the candy furnished to the Army, lemon drops were the soldiers' favorite. Produced from a standard recipe that the Army distributed among its contractors, the Army lemon drop was extra sour, and the wartime troops consumed them at a rate of two hundred thousand pounds a month. The Army's insistence that its candies be individually wrapped, first in a layer of waxed paper, then in a piece of tinfoil, gave added impetus to the movement toward packaging.[26]

The consumption of candy skyrocketed during the war, but more important, the war seemed to effect a permanent change in American eating habits. Per capita consumption of candy rose from 5.6 pounds in 1914 to 13.1 pounds in 1919. The observation of one industry history after World War II, that "the male member of the family has learned through military service . . . many of the useful values of candy," could more accurately be applied to World War I.[27]

Yet the postwar candy habit was expressed not in a newly kindled appreciation for old-fashioned lemon drops but in a passion for new chocolate products. Out of World War I came America's enduring love for the chocolate bar and its kissing cousin, the chocolate-coated candy bar. The rising popularity of chocolate was a consequence not only of its appealing taste but also of a modernizing industry, a growing consumer culture, and an expanding nation at the turn of the century.[28]

Chocolate was processed in America as early as 1765 by James Baker near Dorchester, Massachusetts, yet both production and consumption remained small until the late nineteenth century. In 1880 only seven firms made chocolate in the United States, but twenty years later the Census Bureau tallied twenty-four establishments producing a combined total of nearly $1 million in chocolate. By 1909 the number of firms engaged in chocolate manufacture exceeded one hundred.[29] Such an increase was supported by a determined advertising campaign. In 1906 one industry observer reported:

> Fortunes have been spent and tremendous fortunes reared up in the development of the chocolate business of recent years. The item of advertising alone has become a serious one to the chocolate manufacturer. With the commercial exploitation of "milk chocolates" of foreign and home manufacture there has been a great rush to advertising, and notable campaigns of publicity have been carried on by some of the manufacturers in the newspapers and magazines and out of doors on the billboards. This has resulted in increased sales, and many people have formed the "chocolate habit" who before were not frequent eaters of the dainty.[30]

Escalating chocolate sales were clearly linked to advertising, but the increased availability of chocolate also reflected the larger reality of a nation that was expanding abroad. Chocolate manufacture inevitably depended on overseas trade, primarily with Latin America and Africa, for the importation of cocoa beans.[31] Such trade was facilitated by the United States' territorial expansion following the Spanish-American War. In the immediate aftermath of the war the industry sometimes saw the new U.S. territorial possessions primarily as markets; for example, one industry journal boasted that three caramel factories in Pennsylvania produced "something like 15 caramels for each inhabitant of the United States, Porto Rico, Hawaii and the Philippines."[32] At other times it believed these hot tropical climates would be more important as sources of raw materials. Thus the *Confectioners' and Bakers' Gazette* reported that "all the insular possessions of the United States are cacao-growing countries, and the product, characteristically American, is obtaining, as is its due, a more and more important place in national dietetics."[33] In a finely intertwined economy the territorial expansion that solidified the U.S. connection to sources of cocoa relied upon an army that provided an important market for chocolate products. Indeed, military provisioning was the midwife to a new mass-market candy, the chocolate bar.

Until World War I chocolate was marketed in a variety of forms. Assorted chocolates, candy centers dipped in chocolate, claimed the upper price levels. Less expensive were the penny candy "cigars" favored by children. For the adults there were small bags of penny-weight milk chocolate, often mixed with peanuts, which the retailer cut off larger planks on request. To avoid the inconvenience of cutting individual portions by hand, a few pioneering candy companies sold packaged candy bars consisting of almond nougat or chocolate-coated marshmallows with peanuts as early as 1911. These packaged candies were particularly popular with the ballpark trade, where the convenience of its triple-seal package had already ensconced Cracker Jack in the cultural lore of the baseball experience.[34]

The same lure of convenient, packaged food prompted the army to issue one-ounce chocolate-sugar cakes as emergency rations during World War I. They were so popular with the troops that many soldiers ate them as soon as they were issued, ignoring the army's admonition that they were special-occasion provisions. That wartime introduction to the packaged chocolate bar produced a generation of willing consumers, a coterie of young men who bought a profitable line of novelty candy bars in the postwar era.

Such packaging was a necessary first step to the introduction of brand-name products. Susan Strasser argues that the appearance of nationally advertised, trademarked brands was an essential element in the rise of new mass markets. They replaced bulk goods sold by local grocers and established a direct relationship between consumers and manufacturers. Consumers learned to rely upon the reputation of national manufacturers rather than local retailers to ensure the quality of their purchases. A trademarked brand carried the manufacturer's name prominently on the label as assurance that the firm stood behind its product.[35]

The effort manufacturers expended in establishing the reputation of their products allowed them to compete on the basis of reputation rather than price. For example, both eastern and western manufacturers agreed that the quality of their candy was equal, but eastern candies fetched eighty cents a pound in California, whereas western candies could command only fifty cents there. As one manufacturer acknowledged, "Eastern makers have the reputation and they get the prices." Packaging, branding, and advertising were essential elements, then, in breaking free of the price competition that had so concerned Page.[36]

With the individually wrapped candy bar, candy in general and chocolate in particular broke free from its traditional identity as the marker of a special occasion to become an everyday indulgence. By 1957 one-third of all candy was purchased in this form. While federal Treasury officials continued to burden candy with luxury taxes off and on until the advent of the income tax, the American public was ready to accept candy as an everyday food. After World War I that candy was most likely to be machine-made, packaged, branded, advertised, and chocolate.[37]

Women and the Confectionery Industry

Although the candy industry nicely illustrates the contention that the parallel developments of mechanized production and new merchandising strategies were in fact inextricably linked, the history presented above is only a partial account of the emergence of a mass-production confectionery industry. What of the initial premise that these historical actors lived in a gendered world?

It is not hard to find gender in this spare account of industrial development. The history of mechanization is often written as the loss of men's skilled jobs to automatic machinery, and the displacement of sixty-five candymakers with one machine is yet another example. And whether people ate candy to celebrate a special occasion or as a common treat, the candy consumer was often assumed to be male. The courtship chocolates, the candy cigar, the army lemon drop, the ballpark Cracker Jack—all were marketed to American men. It is not hard to find gender in our industrial histories, for men are often an integral part of the historical model; however, it is difficult to find women.

The invisibility of women in industrial history stems partly from their contemporaries' ambivalence about women's participation in both the work force and the consumer economy. Gender ideology insisted that men should provide the financial support for the family and women should accede to an economically dependent position. If in reality women formed a substantial and growing part of the industrial work force, social theorists hoped they would not be there long. They envisioned a few years only between school and marriage when young, single women would live at home and work in a shop, office, or factory. So in a precarious balance of ideology and practicality employers counted on both women's presence in the work force and their speedy retirement from it, crafting employment policies that incorporated social prescriptions and self-interest.

The assumption that women were economically dependent upon men had an inevitable impact upon their role as consumers as well as workers. With a restricted access to wages, women were uncertain consumers. Young working women who lived at home in conformity with social expectations often bargained with their families for a share of their own wages. Some retained their right to all of the money they earned; others turned over their entire paycheck to their family and received only carfare in return. Those who lived independently found that a woman's wage seldom covered necessities, let alone luxuries. Thus women often relied upon men's gifts to gain access to the commercial economy. Kathy Peiss's portrait of working-class women who relied on their dates to pay their way into the new commercial entertainments, such as public dance halls, amusement parks, excursions, and movies clearly shows the social price single women paid for that lack of independence.[38]

Women's dependence meant that although men might be the buyers of candy, it was often women who were the real consumers. The distinction between who spent the money and who ate the sweets is most easily seen in courting couples. Women usually shared a gift of candy with their suitor on presentation, but the sweets were theirs to finish. The complex meanings that women attached to courtship candy are illustrated by a vignette that appeared in a candy industry journal. It purports to be a conversation on the street between young women, but whether authentic or apocryphal, it captures the nuances of the buyer-consumer dichotomy.

Two young women are talking, one of them holding a box of chocolate nougats. Her girlfriend asks, "Where did you get them? Was it Charley or Bob?"

"It was Charley," she replies. "He always has the nicest candy."

"But you like Bob best," her friend asserts.

The courted woman replies, ambivalently, "Agnes, dear, you are entirely too inquisitive."

"You mustn't get married if you like candy," Agnes counsels her girlfriend. Once married, she says, men seldom buy their wives candy.

Alarmed by this prospect, the courted woman determines to ask their newly married friend, Maude, whether it is true. Maude confirms that her husband no longer buys her candy but hastens to add, "He doesn't buy any candy because I don't want him to. What's his is mine, and I just take his money and buy it whenever I want it. It is a much better arrangement, too, for I buy the kind I like and he seldom did."

The apparent contrast between married Maude and her single friends is between women with money and women without. Women's dependence on men's earning power to gain them access to the delights of the commercial economy is here seen as most acute during the courtship period, when Maude and her friends can only hope their beaux will bring the preferred kind of treats. However, Maude's example of free access to a husband's earnings only obscures a married woman's dependence. This bittersweet candy exchange illustrates the reality that both Maude and the courted woman were economically dependent upon men. Without access to money women might predictably remain consumers rather than buyers.

Yet some women distinguished themselves as enthusiastic buyers of candy. In a clear indication that candy was becoming an everyday indulgence the *San Francisco Chronicle* estimated that the women in that city consumed five tons of candy a day. One saleswoman reported that "'the candy fiends' come to the store every day," and one such "candy fiend" confirmed a passion for her daily ration. She confided that her selection depended on her reserves of time and money: "Sometimes I select the candy that looks prettiest; sometimes that which is the most convenient, and again the cheaper grade—in the latter case you may depend upon it that I have about cash enough to buy lunch and pay car fare." Indeed, the newspaper suggested that women were far more sophisticated purchasers of candy than "the average man whose knowledge of candy is confined to his courting days." The *Chronicle* is silent on the social class of the candy fiends in general, but the mention of daily carfare and lunch money suggests that this woman was a wage worker who used her own money to buy exactly the candy she wanted.[39]

However, women's wages were consistently low compared with men's earnings, especially low in light of the cost of living. Most working women faced the hard fact that their wages seldom covered basic living expenses, let alone such indulgences as candy. Employers justified women's low wages with the argument that their employees were part of a family unit and that since they had fathers or husbands to support them, they had no need for a living wage. This argument both presumed and enforced women's dependent status. Its inexorable, circular logic guaranteed lower labor costs for manufacturers while maintaining women's proper place in the domestic sphere. This view of women's social role was pervasive; even industrial reformers often justified protective legislation for women on the basis of protecting women's primary role as potential wives and mothers.

These general characteristics of women's industrial participation were glaring in the confectionery industry. The low wages, long hours, and frequent unemployment of the confectionery industry led critics to target this industry as one of the most in need of reform. In 1913 the New York State Factory Commission, concerned with the "advisability of fixing minimum rates of wages," investigated four industries notorious for low wages, including the manufacture of confectionery.[40]

The Factory Commission soon discovered that more than 60 percent of confectionery workers were women. Roughly calculated, 60 percent of those women workers were native-born, 60 percent were under 21 years of age, and more than 60 percent earned less than $6.50 per week when employed. Studies also revealed that 75 percent of the women candy workers were single and that employers justified their low wages with the argument that they lived at home and so did not need a subsistence wage. Reformers emphasized the self-fulfilling nature of these assumptions pointing out that only dependent women would be comfortable taking a job at wages below subsistence levels. One investigator, however, questioned the reality of their dependent status; she said that women knew what their bosses wanted to hear and that "most of the girls . . . report that they live at home because they probably won't get the job if they report otherwise."[41] Indeed, one employer was emphatic that unless she lived at home, no girl could live on twelve or fourteen dollars a week and maintain her moral standards. Employers' assumptions that independent women were more likely to turn to vice or crime to make up for wages below the subsistence level did not encourage applicants to be truthful about their living arrangements.[42]

For those who were not supported by fathers or husbands confectionery work provided a marginal existence. One factory worker explained that her forewoman, May, was always grumpy because she was making only one dollar more than the beginners and "she hasn't any parents and has to keep herself on it." Thus feminists argued not simply for better wages but for a living wage that would allow women workers to maintain themselves independent of fathers and husbands.[43]

Female confectionery workers suffered not only from low wage rates but also from seasonal unemployment. While manufacturers fretted over reconciling seasonal consumption with mechanization, workers worried how to reconcile fluctuating production schedules with a decent standard of living. Candy manufacturers met their peak demands by hiring temporary workers at low wages and working their machines and their employees overtime. During the 1905 Christmas rush confectionery workers at one firm began work at 7:15 A.M. each day and logged a seventy-five-hour work week. Long hours prevailed from the beginning of the holiday rush in September until manufacturers reduced their work force in mid-December. Six weeks before Easter some women were rehired as manufacturers prepared for the springtime peak, but summer brought a 45 percent reduction in the work force. Seasonal production was only possible because a vast army of female workers underwrote manufacturers' flexibility. One eleven-year veteran told investigators resignedly, "In a candy factory one is always laid off for a couple of weeks after Christmas and of course in summer."[44]

Although investigators estimated that 60 percent of female confectionery workers were employed fewer than eight weeks a year, that figure probably underrepresented women's employment. Gathered from factory records, these statistics captured women's brief tenure in any one job but failed to report their continued employment in the industry in general. Factory records camouflaged the fact that women often left an unsatisfactory job at the end of the weekly pay period in order to take a position with another firm with the hope of better pay or better working conditions. Indeed, one factory reported a turnover of 300–400 percent during the Christmas rush, when new jobs were easy to secure. As a consequence of this restless search, female confectionery workers often labored for years at "beginner's" wages in an endless round of new but otherwise identical jobs.[45]

Of the women confectionery workers only the highly skilled hand dippers escaped the cycle of low wages and debilitating unemployment. They represented a small fraction of the women candy workers: 53 percent of women worked as packers and wrappers, 25 percent as helpers, 10 percent as supervisors, and only about 13 percent as hand dippers. Whereas packers and wrappers received a median wage of only six dollars a week, hand dippers commanded a median wage of eight dollars. "The dippers are the only ones in this business that can make a living," one woman told an investigator. "Some of the speedy piece workers make good money at this time of year but they lose out in the summer when things are slack."[46]

Hand dipping was skilled work that consisted of coating candy centers with chocolate. One observer described their work as follows:

> The girls work in groups of four clustered around large stone pots, into which the liquid chocolate is poured. Around this pot is a collar, with large flat spaces at its four corners, thereby giving working-room to the four operators or "dippers," as they are called in the factory. To

> these girls the candies or inner portions of the confections are supplied in shallow wooden trays . . . and piled at places within convenient reach of the operators. The latter takes the candy out of the tray, a piece at a time, and with her hand dips it in some of the liquid chocolate which she has spread on the corner of the collar before her. With her fingers she rubs the chocolate over the candy, smoothes it and places it on a tin tray at her side. . . . Each operator ceaselessly stirs [the chocolate] on the corner in front of her with her right hand while she feeds it with the candies to be covered with the left.[47]

Paid by the piece, a good dipper could turn out a hundred pounds of chocolates a day. High wages made their jobs the most desirable, one woman explained, so "the brighter girls begin as wrappers and learn to be fancy packers or try to become dippers."[48] These women embodied what reformers hoped to achieve for the majority of candy workers.

Yet the hand dippers' privileged position was relative only to their sisters'. Male machine operators earned a great deal more than female hand dippers. In 1913 a wage of eleven dollars a week marked the divide between men's and women's wages; 90 percent of male machine operators made eleven dollars a week or more, whereas 96 percent of hand dippers earned less than eleven dollars a week. Like most factory work in this period, production work was gendered; pay scales for men and women were just as separate as the kinds of work to which they were assigned. One investigatory group noted, "Men do the machine work; women are employed to clean and wrap the cakes."[49]

The strict separation of men's and women's jobs allowed employers to justify wages in terms of the skill required to accomplish each task. "In general, it may be said that cooking or 'making' candy is a skilled trade in the hands of men. Machine tending which requires judgement is also a male occupation," explained one authority. On the other hand, "packing and wrapping require deftness and an eye for effect, which have made them distinctly women's lines." This division of industrial work, relying upon gender stereotypes about men's and women's abilities, established two distinct types of work that were separate and unequal. Women were paid a woman's wage first and then a skilled wage within the limits of that range.[50]

Given the gendered nature of work in candy production, it is not surprising that mechanization of the industry affected men and women differently. By 1900 a machine called the "mogul" was introduced into U.S. factories to mechanize the production of candy centers. It integrated three steps of production by printing the shape of the candy in a bed of starch to form a mold, filling the mold, and then freeing the completed center from the starch. The mogul was followed three years later by the "enrober," which encased the centers in a chocolate coating. Page's Confectioners' Machinery and Manufacturing Company marketed the machines as a means for eliminating the problems of all too human employees and decreasing the cost of skilled labor. The mogul nibbled away at men's jobs, and the enrober threatened female hand dippers.[51]

However, not all companies promoted the deskilling of their workers.[52] Hand dippers were ununionized, highly skilled workers secured at a bargain wage—a woman's wage. Manufacturers might think twice about the economic benefits of mechanizing these jobs. Women workers labored for a bargain wage and yet were tied to a high-price markets. Indeed, hand-dipped chocolates brought high prices on the market because consumers associated them with high quality. In 1920 *Scientific American* explained that "the most expensive candies, those which retail at two dollars a pound and upwards, are still made almost entirely by hand." Hand-dipped candies used a more expensive grade of chocolate, but their appeal derived not simply from their taste but from their hand-crafted quality. Women's labor had a special cachet with the consumer; as *Scientific American* explained, hand dippers "in general follow the same sort of technique that sister uses" in the kitchen at home. Such positive associations with home production connected women workers with a high-priced market. The public valued not just handmade goods but goods made by women, which they imbued with all the virtues of the domestic sphere.[53]

Contrary to a simple model of mechanization in which machines readily replaced expensive skilled labor, some confectionery manufacturers expanded their output of hand-dipped candies during the height of mechanization. Manufacturers found that the savings that resulted from adopting the mogul to form the candy centers allowed them to profitably hand dip cheaper grades of candy; one large Philadelphia firm began hand dipping even fourteen-cent chocolates. A well-known Philadelphia manufacturer went on record with the prediction that "hand manipulation would supersede machinery in the near future." Mechanization of women's work followed a different path than that of men's, for their gendered wages tied them to different consumer markets. So hand dippers survived mechanization and provided an exemplary model of the woman wage earner in the confectionery industry.[54]

Reforming the Confectionery Industry

Just as women's presence in the work force shaped the pattern of candy production, their influence in the marketplace was undeniable. Although few single working women earned much in wages, women in general commanded considerable buying power. Prosperous women sometimes enjoyed a substantial amount of discretionary spending, whereas housewives who handled the family budget controlled a smaller amount of money with a great deal of discretion. Beginning in the late 1890s some elite and middle-class women in Boston and New York banded together in consumers' leagues to use their economic power for reform. They were less interested in protecting buyers from shoddy or defective products than in improving conditions for workers. Mrs. Charles Russell Lowell, founder of the movement, insisted that it was "the duty of consumers to find out under what conditions the articles they purchase are produced and distributed, and to insist that these conditions shall be wholesome and consistent with a respectable existence on the part of the workers."[55]

The consumers' league movement was a feminist undertaking that linked not simply consumers and workers but women consumers and women workers. American women, long denied the vote and full citizenship, had a legacy of using consumer power to achieve political goals. The politicization of consumption dates to the American Revolution, when the substitution of homespun cloth for British-made fabrics assured colonial women a patriotic role if not a real voice in the political crisis. Similarly, Sarah Pugh, a radical Philadelphia Quaker abolitionist, refused to use sugar and cotton because they were the products of slave labor. Thus, consumer power had long been a woman's political tool, adaptable either to the individual conscience or the group endeavor without the need of government sanction.

The most active of the consumer groups was the National Consumers' League (NCL), organized in 1899 when four established local leagues combined to form a national organization. It was not an exclusively female organization; indeed, Newton D. Baker served as its founding president. But the active participation of women in the leadership of the local organizations was capped by the appointment of Florence Kelley as secretary of the national organization. Kelley brought with her a keen interest in the problems of women and children, which she continued to pursue through the organization. In its techniques, its leadership, and its clientele, then, the NCL had a distinctly feminist cast.[56]

The first beneficiaries of consumer-led reform were shopgirls who worked long hours during the Christmas rush. "Shop Early" campaigns rallied consumers behind the league's demand for shorter work hours. Soon the reformers grew concerned about factory workers as well, and they proposed awarding an NCL label to manufacturers whose wages and conditions met stringent standards. With the NCL label as their guide, shoppers could support conscientious businesses and boycott those that refused to meet league demands. The modern strategy of marketing packaged, branded, and advertised goods now threatened to work against producers. Reformers understood that large corporate manufacturers were increasingly common, "and with [them] the widely advertised trade name."[57] They used the accountability of brand-name products to force manufacturers to adopt higher wages and better working conditions.

When the NCL espoused industrial reform on behalf of confectionery workers, manufacturers' heightened concern about their reputations provided the league with an important lever for reform. Candymakers had discovered that selling packaged, brand-name products allowed them to compete on the basis of the quality and reputation of their goods rather than the price; they had never imagined that the public would concern itself with production. Yet, the secretary of the New York Consumers' League (NYCL) explained, "The candy manufacturer has spent thousands of dollars to make his name known to the consumer; he has reason to fear lest the stigma of low standards be attached to that expensively famous name." Reformers used the linkage between producers and consumers—originally strengthened by manufacturers to more closely match consumption to factory production—to their own ends.[58]

The use of consumer power to achieve reform in the workplace was successful particularly in the early years. At times the NCL relied mainly on governmental regulation to improve the factory workplace, but when the Supreme Court overturned minimum-wage laws in 1923 the consumers' leagues turned again to "moral suasion" and the consumer boycott. So in 1928 the NYCL acted on a long series of investigations into conditions of employment in the confectionery industry by mounting a consumer assault on long hours, low wages, and seasonal employment. They assembled statistical data on women in the candy factories and supplemented them with the personal account of an investigator who had taken jobs in twenty-five different candy factories as part of her investigation. The result, a report entitled *Behind the Scenes in Chocolate Factories*, galvanized both consumers and candy manufacturers.

In the aftermath of the report one consumer wrote to the NYCL, "I see by the newspapers that you have made an investigation of the cleanliness of candy factories in New York City. Will you kindly send a list of those you found satisfactory as to cleanliness, so that my friends and I know at what places to buy our sweetmeats?" And another anxious shopper reported, "I have just read your booklet on the candy industry and am sure that from now on, unless I can find out which are the pure candies made under sanitary conditions, I can't buy any more candy." Within a week fifty-seven factory owners capitulated to the reformers' demands; of those, only ten had initially met all standards. Indeed, the NYCL was surprised by the extent of its success, and the executive secretary reported a strain on the league's finances as it hastily hired investigators to meet manufacturers' demands for immediate inspection. Most candymakers quickly accepted the league's standards of cleanliness, yet it was only after considerable insistence that they agreed to bring the minimum wage up to fourteen dollars a week. Upon compliance they were placed upon a candy white list, which assured consumers that they had complied with minimum wage and sanitary conditions.[59]

However, by 1928 consumerism was becoming more complex. The consumers' leagues of the Progressive era were originally rooted in a heartfelt collective interest based upon sisterhood that cut across classes. Yet later consumer groups increasingly held their membership together with an appeal to rational self-interest. During the drive to establish the candy white lists consumerism was teetering on the edge between these two formulations of consumer power, one "gendered" and the other "rational." Experts such as government bureaucrats, home economists, and advertising executives each promoted their own brand of rational consumption. Without an explicit political ideology holding workers and consumers together, by 1928 consumers tended to behave as the creatures that advertising executives urged them to be—discriminating rather than political.

Nowhere is the conflict between the self-interest of consumers and the larger goals of feminist reform clearer than in the friction over the issue of sanitation. The 1928 candy investigations railed against low wages and poor working conditions, but the consuming public fastened its attention more narrowly on the report's account of unsanitary conditions. Hand dipping drew its share of critical attention, and the report proclaimed the enrober "more sanitary." The effect was to encourage a consumer distaste for handmade goods, undermining the positive associations of the commercial hand dipper with "Sis" in the kitchen at home. This assault on the hand dippers' market niche clearly placed consumers' interests before preserving those few skilled jobs for women. Hand-dipped chocolates continued to be marketed, but the number of hand dippers diminished rather than grew. When the Depression made its own assault on the market for high-priced chocolates hand dippers could be found only in specialty shops. Machine-made chocolates now had their own appeal.[60]

Conclusion

Thus, behind the chocolate bonbon is a more complex story about the links between production and consumption than the traditional story behind the Model T. It casts a critical light on the assumption that mass production finds its essential connection to the consumer through lowered prices, the claim that the adoption of mass production provides economies of scale that both increase manufacturers' profits and decrease consumers' costs. In the confectionery industry mechanization was linked instead to new methods of merchandising that freed manufacturers from price competition. Because manufacturers recognized the consumer appeal of intangible factors such as a woman's touch, mechanization itself became a complicated affair. Looking past an old emphasis on price competition, we can see the complexity of the bonds that tied producers and consumers together.

One of those ties was gender. Manufacturers easily recognized the distinctive patterns that gender roles produced. They viewed their workers as candymakers and hand dippers and their customers as soldiers and valentines. That firms manufactured and advertised products aimed specifically at men and women is not remarkable, but that consumers created their own conception of gender identity through consumption should not be overlooked. Women who bought candy with their own money confounded the ideal of the dependent female, and those who wielded consumer power for industrial reform repudiated it altogether. Manufacturers intended brand-name products to protect their economic interests, but power flowed two ways in the new producer-consumer network.

Notes

1. "Development of the Candy Industry," *Confectioners' and Bakers' Gazette* 25 (Dec. 1903): 28.

2. On the interrelationship between the American system of manufactures, quantity production, and mass production see David A. Hounshell, *From the American System to Mass Production, 1800–1932: The Development of Manufacturing Technology in the United States* (Baltimore: Johns Hopkins Univ. Press, 1983). A large literature modifies our traditional understanding of mass production as synonymous with Fordism. This more complicated picture includes flexible production, batch production, and other variations (see Philip Scranton, *Figured Tapestry: Production, Markets, and Power in Philadelphia Textiles, 1885–1941* [New York: Cambridge Univ. Press, 1989]).

3. Roland Marchand, *Advertising the American Dream: Making Way for Modernity, 1920-1940* (Berkeley: Univ. of California Press, 1985); Stephen Meyer, *The Five Dollar Day: Labor Management and Social Control in the Ford Motor Company, 1908–1921* (Albany: State Univ. of New York Press, 1981); Susan Strasser, *Satisfaction Guaranteed: The Making of the Mass Market* (New York: Pantheon, 1989).

4. In 1932 the Mexican painter Diego Rivera painted a mural of the Ford Company's River Rouge plant on the walls of the Detroit Institute of Art. Ford's River Rouge plant, located in Dearborn, Michigan, was famous for its scale and its integrated production methods. Images of men behind the wheel of an American

car were often used in advertising; it seems likely that the famous advertising firm of Batten, Barton, Durstine and Osborn (BBD&O) created such an image.

5. *Confectioners' and Bakers' Gazette* 28 (Dec. 1906): 12.

6. "Lasses Candy by Machinery," *Confectioner and Baker* 6 (Jan. 1901): 5 (old series vol. 24, no. 261).

7. "Confectioners' Machinery Company Given World's Fair Award," *Confectioners' and Bakers' Gazette* 26 (Nov. 1904): 30.

8. Vivian M. Moses, "A Newspaper on the Chocolate Trade," ibid. 27 (May 1906): 24.

9. *Confectioners' and Bakers' Gazette* 28 (1906): 12. Advertisement by H. C. Hawk, ibid. 26 (May 1905): 34.

10. Prices declined partly as a result of a decrease in the cost of labor but also because of a reduction in the cost of ingredients. The price of sugar dropped, and the value of beet sugar production rose from approximately $283,000 in 1880 to more than $7 million in 1900. In addition, the use of glucose as a sweetener further reduced expenses; when the National Academy of Sciences issued a report around 1887 that affirmed the healthfulness of glucose, use of that sweetener grew quickly ("Development of the Candy Industry," 28).

11. D. Auerbach, "The Nation's Candy Bill," *Confectioners' and Bakers' Gazette* 31 (Oct. 1909): 26. Despite Auerbach's claim, many types of confectionery machinery originated in Europe: the revolving steam pan was imported by "two Frenchmen" in 1851; the enrober was invented in France and imported around 1900; an Austrian device for continuous vacuum cooking was introduced to the United States in 1906. These devices revolutionized production in pan goods, chocolates, and hard candies, respectively (see Philip P. Gott and L. F. Van Houten, *All about Candy and Chocolate: A Comprehensive Study of the Candy and Chocolate Industries* [Chicago: National Confectioners Association of the United States, 1958], 16, 22). In chocolate processing, the development of machinery in nineteenth-century Europe included Philippe Suchard's melangeur (1826), Coenraad Van Houten's cocoa press (1828), and Rudolphe Lindt's conche (1879). American chocolate manufacturers often used imported equipment, such as Milton Hershey's first melangeur, made by Lehmann of Dresden, or machines manufactured under foreign license (see Sophie D. Coe and Michael D. Coe, *The True History of Chocolate* [London: Thames & Hudson, 1996]; and Chantel Coady, *Chocolate, the Food of the Gods* [San Francisco: Chronicle Books, 1993]). Indeed, one industry writer reported that foreign dealers in confectionery machinery all but controlled the American market in some lines despite a 45 percent duty (see "Confectioners' Machinery Company Given World's Fair Award," 30).

12. Frederick Klinck, "The Candy Store of Thirty Years Ago," *Confectioners' and Bakers' Gazette* 28 (Dec. 1906): 22; U.S. Department of Labor, Women's Bureau,

Women in the Candy Industry in Chicago and St. Louis: A Study of Hours, Wages, and Working Conditions in 1920–21, Bulletin no. 25 (Washington DC: GPO, 1923), 1.

13. Quotation from the *San Francisco Chronicle,* reprinted in "The Trade in San Francisco," *Confectioners' and Bakers' Gazette* 26 (June 1905): 27; "Confectionery Manufacturers and the New England Confectionery Company," ibid. 32 (Mar. 1911): 30.

14. Price competition was a constant complaint of the industry, and particularly in the years before the Pure Food and Drug Act of 1906 it led to adulteration of candies with cheaper ingredients. The first trade association, the United States Manufacturing Confectioners Association, established in 1876, tried to establish a scale of prices and standard terms of sale. However, the myriad small producers made it impossible to achieve conformity, and the organization collapsed. Individual manufacturers responded by maintaining the quality of their primary line of candies and also selling lower-quality "Number 2" goods. A second trade association, the National Confectioners' Association of the United States, established in 1884, vigorously promoted antiadulteration laws that forbade the use of harmful additives such as terra alba. However, the association was less successful in persuading its members to give up Number 2 products, which more than likely used harmless extenders such as flour or cereal. These grades were a necessary part of competing for business (see Gott and Van Houten, *All about Candy and Chocolate,* 163–96).

15. F. H. Neely, "Arrangement of Machinery." *Confectioners' and Bakers' Gazette* 30 (Jan. 1909): 15.

16. "Confectioners to Own Machinery Plant," ibid. 31 (Oct. 1909): 26. The article "Confectioners' Machinery and Manufacturing Company to Add to Its Plant," ibid. 27 (July 1906): 25, gives the date as 1900.

17. Frank H. Page, "Something Different," ibid. 27 (Dec. 1905): 17.

18. Manufacturers who wanted to sell fresh candies could not produce their goods too far in advance of sales. Studies following World War II suggested that vegetable fats developed a rancid taste within six to twelve weeks if stored at a warm 86°; animal fats were even less stable. By custom few firms in the early decades of the twentieth century began Christmas production before early September.

19. Sidney Mintz, *Sweetness and Power: The Place of Sugar in Modern History* (New York: Penguin, 1985), 122, 151.

20. *Confectioners' Gazette* 35 (Feb. 1914): 14.

21. Gott and Van Houten, *All about Candy and Chocolate,* 20; E. C. Beynon, "Easter Confectionery Store Window Displays," *Confectioners' Gazette* 35 (Mar. 1914): 28.

22. Beynon, "Easter Confectionery Store Window Displays," 14.

23. Gott and Van Houten, *All about Candy and Chocolate,* 164; D. Auerbach, "The Nation's Candy Bill," 26.

24. Mintz, *Sweetness and Power,* 150. From 1880 to 1900 the now familiar cardboard box replaced simpler paper bags, and from 1900 to 1925 the industry enthusiastically adopted the use of glassine, cellophane, and aluminum (Klinck, "The Candy Store of Thirty Years Ago," 22; Gott and Van Houten, *All about Candy and Chocolate,* 26, 120).

25. Rohland A. Isker, "For That Extra Effort," in Gott and Van Houten, *All about Candy and Chocolate,* 97–112; "Development of the Candy Industry," 28; George Bernard Shaw, *Plays by George Bernard Shaw* (New York: Signet, 1960), 124.

26. "Candy in the Army," *Literary Digest* 61 (5 Apr. 1919): 28; Gott and Van Houten, *All and Candy and Chocolate,* 103.

27. U.S. Department of Labor, Women's Bureau, *Women in the Candy Industry in Chicago and St. Louis,* 1; Gott and Van Houten, *All about Candy and Chocolate,* 153–54.

28. In 1957 candy bars accounted for a third of American sales in terms of both weight and value (Gott and Van Houten, *All about Candy and Chocolate,* graph between 112 and 113).

29. "We Make Chocolate," ibid. 31 (Oct. 1909): 27; "Development of the Candy Industry," 28.

30. Moses, "A Newspaper on the Chocolate Trade," 24.

31. Cocoa is a native American plant that prospers best in the equatorial regions, where the temperature is consistently above 60°. The plant was introduced into the Ivory Coast and Ghana in 1879 (see Gott and Van Houten, *All about Candy and Chocolate,* 42).

32. "Caramels for the World," *Confectioner and Baker* 7 (June 1901): 22.

33. "We Make Chocolate," 27.

34. Gott and Van Houten, *All about Candy and Chocolate,* 18, 20. One famous 1908 song contained the lyrics, "Take me out to the ballpark, . . . Buy me some peanuts and Cracker Jack." Cracker Jack was well established before World War I, but it tapped into the patriotism of the public by adding the now familiar sailor boy to its package during that conflict.

35. Strasser, *Satisfaction Guaranteed.*

36. "Trade in San Francisco," 27.

37. Gott and Van Houten, *All about Candy and Chocolate,* graph between 112 and 113.

38. Kathy Peiss, *Cheap Amusements: Working Women and Leisure in Turn-of-the-Century New York* (Philadelphia: Temple Univ. Press, 1985).

39. "Trade in San Francisco," 27.

40. The other three industries were the manufacture of paper boxes, the manufacture of men's shirts, and retail stores.

41. Mary Dewhurst Blakenhorn, "Who Makes Your Candy?" *Survey* 60 (15 Apr. 1928): 115.

42. Ibid.

43. *Behind the Scenes in Chocolate Factories* (New York: Consumers' League of New York, 1928), 11. At its eighth annual meeting, the National Consumers' League voted to undertake a study of wages and the standard of living of self-supporting women. Out of this study came the publication *Making Both Ends Meet* (see *The National Consumers' League: First Quarter Century, 1899–1924* [New York: National Consumers' League, (1924)]: 12).

44. Mary Van Kleeck, *Working Hours of Women in Factories* (New York: Charity Organization Society for the National Consumers' League, 6 Oct. 1906): 10; U.S. Department of Labor, Woman in Industry Service, *Wages of Candy Makers in Philadelphia in 1919*, Bulletin no. 4 (Washington DC: GPO, 1919): 30.

45. Blakenhorn, "Who Makes Your Candy?" 115.

46. *Behind the Scenes*, 11.

47. Moses, "A Newspaper on the Chocolate Trade," 25–26.

48. New York State Factory Commission, *Confectionery in New York City* (New York, 1913), 76–77.

49. Ibid., 71.

50. Ibid.

51. Page's company was founded in 1893 on the basis of his patents. It was reported that the company had its "own patents" on two machines, the mogul and the enrober, that were considered "world standards." In addition, however, it maintained an alliance with the celebrated Parisian firm A. Savy, Jeanjean & Co. ("Confectioners' Machinery and Manufacturing Company to Add to Its Plant," 25). Introduced to American around 1900, wooden moguls were replaced by steel ones after 1914.

52. For a discussion of this see the work of three scholars presented at the 1993 annual meeting of the Society for the History of Technology, in Washington DC, in a session called "Keeping Workers in the System: American Managerial Resistance to Labor-saving Innovation, 1880–1920." Papers in that session included Jack Brown, "Custom Products and a Reservoir of Skill at the Baldwin Locomotive Works, 1890–1911"; Kenneth Lipartito, "When Women Were Switches: Technology, Labor, and Female Telephone Operators, 1890–1920"; and Steven Usselman, "Mixed Signals: The Annoying Allure of Automatic Train Control for American Railroads." Lipartito's paper was published in the *American Historical Review* 99 (Oct. 1994): 1074–1111.

53. H. A. Mount, "Uncle Samuel's Sweet Tooth," *Scientific American* 123 (24 July 1920): 88.

54. "Development of the Candy Industry," 28.

55. *The National Consumers' League: First Quarter Century*, 1.

56. The original member organizations were local leagues from New York, Pennsylvania, Massachusetts, and Illinois.

57. Blankenhorn, "Who Makes Your Candy?" 116.

58. Ibid.

59. Eleanore Von Eltz, "The Saints from the Sinners," *Survey* 60 (15 May 1928): 234–35, quotations on 234. See also "The Candy Cleanup," *New Republic*, 11 July 1928, 202.

60. *Behind the Scenes*, 19. In *The Americans: The Democratic Experience* (New York: Vintage, 1973), 89–164, Daniel Boorstin claims that consumer acceptance of mass-produced goods in the United States was rooted in the nation's democratic character. However, that obscures the process by which the public attached values to the terms *machine-made* and *handmade* and the extent to which the definition of those values was a contested issue.

Four

Getting Housewives the Electric Message

Gender and Energy Marketing in the Early Twentieth Century

James C. Williams

Introduction

Between 1910 and 1930 the electric power industry reached into homes across the nation. An array of new household tools appeared on the market whose introduction into the home altered many domestic tasks, mechanizing and otherwise helping to transform home life. Because the electric power industry required a substantial outlay in capital equipment, utilities had to cultivate electric power consumption in a variety of ways if they were to make profitable and efficient use of their investments. This was particularly true in the underdeveloped and thinly populated southern California region, where utilities grew very rapidly and quickly encumbered themselves with excess generating power. Consequently, without prior design or planning, utilities took on the unfamiliar task of encouraging adoption and use of new domestic technologies.

Gender made a difference in the marketing of domestic electric power. The industry's male engineering and sales staffs initially took full responsibility for devising marketing strategies to introduce the new domestic electric technologies to consumers. Their misconceptions about what women actually did and how they functioned inside the home made the utility businessmen ignorant and ineffective in promoting complex elec-

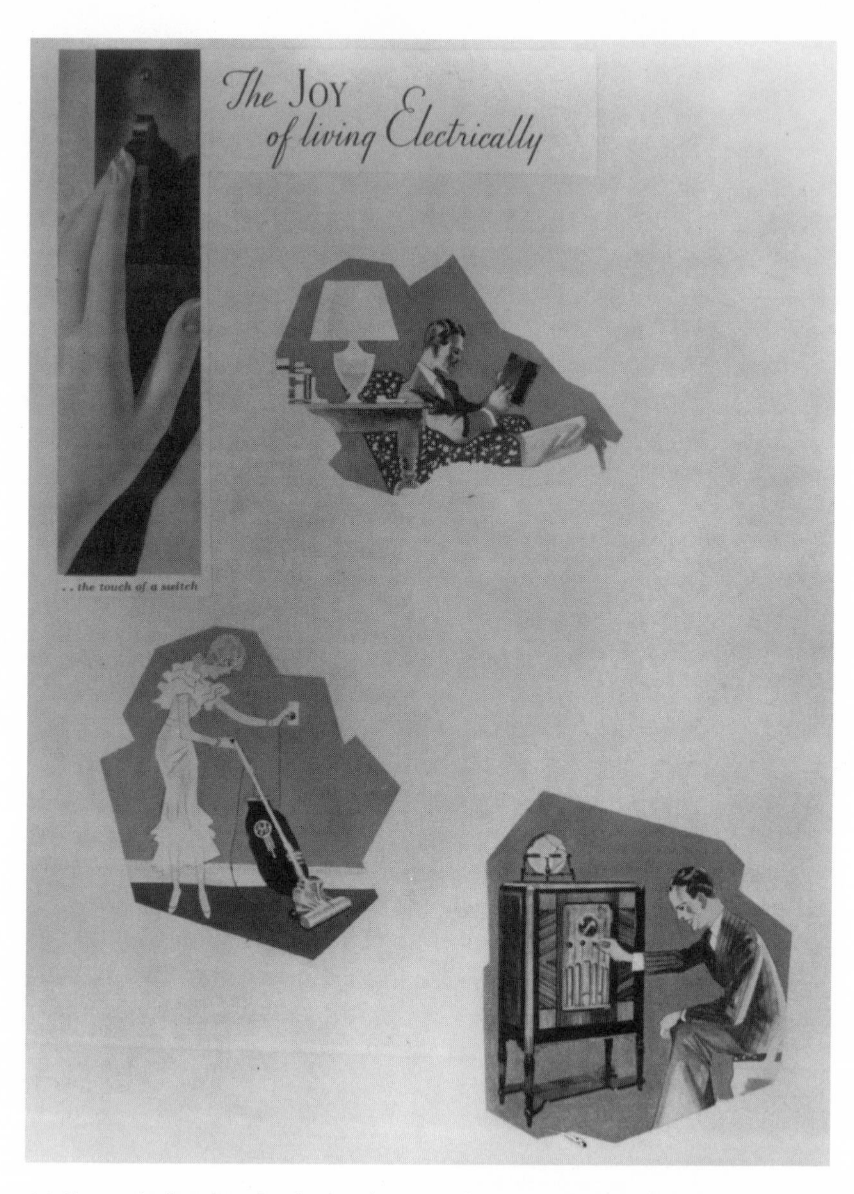

4.1. Domestic electric technology and commonly perceived gender roles in the home. A page of advertising copy from 1934. Courtesy Southern California Edison Company.

tric tools such as ranges and other cooking devices that entailed dramatic changes in household practices. They were more successful in promoting smaller tools, such as electric irons; householders accepted them readily because such tools followed, and clearly made easier, existing patterns of women's work (fig. 4.1).

Successful adoption and use of domestic electric technology and the dynamic of the consumer market required that firms consider of gender, and those that ignored it risked ineffective or failed marketing strategies. Once the electric industry recognized that the housewife was the principal consumer and user of the new household appliances, it increasingly included women in its marketing efforts. During the 1920s utilities feminized their sales approach in order to more effectively target women.[1]

Creating Electric Consumption

At first glance practical economic and technological considerations in the electric power industry seem to have propelled the marketing of household technology. Central-station power producers sought to balance as well as to build load factor, that ratio representing average and maximum power demand during a specified period. Since electricity could not readily be stored, companies needed a generating capacity at least equal to the maximum load or total customer demand. Moreover, whenever steam or hydroelectric power plants stood idle profits were lost. Therefore, to attain the most profitable and technologically efficient operating economies companies sought customers whose low and high periods of power consumption offset each other.[2]

In turn-of-the-century California businessmen eagerly developed and applied electric power. Four decades of gold mining had provided Californians with a rich knowledge of hydraulic engineering. By the 1890s local entrepreneurs and engineers had begun to harness water power to generate electricity, learning from, even using hydraulic works developed by the mining industry. Entrepreneurial electric power companies built scores of hydroelectric plants in the Sierra Nevada, the San Gabriel, and other mountain ranges and mastered high-tension, long-distance electric power transmission so as to deliver electricity to San Francisco, Los Angeles, and smaller distant cities. They dreamed of fostering industrial growth, but initially they found that people wanted their power primarily for nighttime lighting.[3]

Load problems soon bedeviled power companies. In 1912 S. M. Kennedy, the general agent of the Southern California Edison Company (SCE) recalled the plight his firm had faced eight years earlier, in 1904: "The Edison company had water going to waste during the daylight hours; it had big steam turbines standing idle, waiting for a day load." Even though Los Angeles was the fastest-growing metropolitan area in California—its population increased fivefold between 1900 and 1920—and most houses built there after 1900 were wired for electric lighting, SCE's load dissolved during the daytime. In domestic appliances that could be plugged into lighting fixtures, recalled Kennedy, "a large and satisfactory load was ready to be taken, only wanting an invitation . . . the customers needed the appliances, although they did not know it, and the company needed their business and knew it."[4]

Finding the pace of local hardware stores' sales of lamp-socket appliances too slow, SCE undertook aggressive, direct marketing. It opened showrooms at its various offices and advertised the advantages of the appliances in newspapers, circulars, and letters. It loaned electric irons and percolators, sold them on an installment plan, and took old flatirons and coffee pots in trade. It focused on the sale of appliances that were "popular, efficient, durable, and likely to be regularly in use."[5] It purposely avoided pushing sales of ranges, air and water heaters, and other appliances exceeding the industry's accepted limit of six amperes for lamp-socket installations. This permitted the company to load up its residential lines during daylight hours without investing in upgraded transformers, meters, and other services, such as the installation of outlets for electric plugs. Their own convenience took priority over the convenience they claimed to be marketing to householders.[6]

By 1907 the firm had determined that "the only way to sell large quantities [of appliances] and do it quickly was to send expert salesmen to call on customers in their homes." A sales department was organized with fifteen sales representatives, all male. Through regular house-to-house canvassing they recorded each household's appliances, gauged the market, and pushed only those devices that they believed received regular use. In

4.2. Four dapper young salesmen with the tools of their trade in 1910. Courtesy Southern California Edison Company.

1911 SCE reported sales of 15,438 electric irons, 4,634 coffee percolators, 3,440 toasters, and 2,000 other miscellaneous devices. The salesmen soon accounted for 80 percent of the company's total appliance sales, which by 1915 reached 141,705, and SCE achieved a national reputation for its marketing of domestic lamp-socket appliances (fig. 4.2).[7]

Men, Gender, and Marketing

SCE sales staff marketed lamp-socket appliances on the one hand because it served their firm's interests in building load and revenue; but their efforts also intensely reflected the influence of gender. Marketing domestic electric technology occurred at what historian Ruth Schwartz Cowan has called "the consumption junction," the boundary between production and consumption at which technologies are diffused. It is here that agents of technological diffusion introduce more and more people to new technologies, helping them to become conversant with technical processes or devices. The first agents of diffusion in the electric power industry were universally male, and their marketing efforts reflected male characteristics as well as male perceptions of gender.[8]

Men in the utility industry believed in progress through technology, and their very faith was rooted in male identification with technology and the idea of progress.[9] Electricity provided a compelling symbol of cultural modernity. Its rapid adoption wherever inanimate energy could be used was part and parcel of technological advances and the forward march of modern America. It was powerful, efficient, labor-saving, clean, and invisible. It seemed to possess magical qualities that made it an indisputable force for better living, and utility businessmen never conceived that others, men and women alike, would not immediately see this. California electrical engineer Frank Baum articulated what he assumed was all Americans' emotional investment in the promise of electricity when he predicted before the 1915 International Engineering Congress in San Francisco that "some time in the future a nation's civilization will be measured largely in terms of kilowatt-hours consumed per capita." And it did not seem fatuous at all to say, as did a Chicago engineer, that an electric fan or electric iron was "an aid to civilization on a hot day."[10]

California's hydroelectric industry was developed in an unquestionably masculine fashion. Utility entrepreneurs imposed their will over nature and technology through a structured, linear plan, "setting up water wheels in lonely mountain places and making them give light and cheaply turn other wheels in towns miles away."[11] When they began marketing domestic appliances, their approach incorporated this control of technology and of the environment in which they worked. They aimed merchandising strategies to build favorable load factors by increasing domestic power consumption, and a martial spirit flavored their appliance sales campaigns. Historian Mark Rose reports that Denver Gas and Electric Company sales representatives stood morning roll calls, had daily briefings, were checked up on in the field, and were not permitted "to smoke cigarettes in public view."[12] Similarly, SCE salesmen kept detailed daily records, learned sales techniques through lantern slide presentations, and at one point in time wore military-style uniforms. In effect, utility businessmen evinced the sort of "hard mastery" over technology that Sherry Turkle observed in schoolboys learning to use and control computers in the 1980s. This mastery was in stark contrast to the "interactive" or "relational" style, the "soft mastery" Turkle saw in girls working with computers.[13]

SCE's linear and structured approach to building company revenue and daytime load through marketing domestic technologies also was typified in theoretical constructs developed by its engineering and sales staff to show what uses made up load at various times during the day. S. M. Kennedy claimed in 1912, for example, that the most popular appliance sold, the electric iron, was used regularly because the daytime load curve charted for one of SCE's residential districts peaked on Tuesday, the day after the traditional laundry day, declined a bit on Wednesday, and stabilized for Thursday, Friday, and Saturday. In addition, his firm kept track of appliance sales from month to month and year to year. Nevertheless, Kennedy admitted that "some of the strongest evidences of actual use of electric appliances on the residence circuits are to be found in the troubles which are reported when the service on any of such circuits is interrupted."[14] Thus, no matter how accurately their data reflected reality, in the beginning Kennedy and his colleagues were not really sure how effective their efforts were.

At first utility sales representatives were not much better informed than their potential customers about the use of electricity in various settings. Moreover, they often misconstrued what women actually did in the home. Edgar A. Wilcox, a "load builder" and electric-heating specialist with northern California's Great Western Power Company, shed light on these problems in his 1916 book *Electric Heating*.[15] He aimed his work at those who sold electric-heating appliances, offering them basic information about heating with electricity, from measuring temperatures to distinguishing between radiation, conduction, and convection. He sketched out the benefits of electric heat over heat gained from various fuels, provided detailed information on safe wiring, and discussed electric rates. The bulk of his book, however, reviewed household, commercial, and industrial electric appliances and emphasized the modern, progressive qualities of electricity. Wilcox's discussions of both lamp-socket appliances and electric ranges starkly revealed common views of male and female roles in and out of the household.

Wilcox's description of types and models of lamp-socket appliances ranged from southern California's Hotpoint "El Grillo" tabletop grill and "El Bako" oven to chafing dishes and toaster stoves from Simplex, Westinghouse, and General Electric. Collectively he described these appliances as "labor saving devices." Not only had they transformed the industry's economic condition, he said, but they had freed the housewife from drudgery, "making housework an enjoyable pastime."[16] The men who hawked lamp-socket appliances probably never imagined that plugging into light sockets was a specious convenience; however, as Fred E. H. Schroeder noted in his 1986 study of electric plugs and receptacles, "one might plug in an iron at the cost of being left in the dark." Cords twisted when screwing in plugs, short circuits or shock could result from jerking a cord beyond its length, and having to reach lamp-sockets that were placed high up on walls or in the ceiling was at best annoying. On the whole, lamp-socket devices were not as handy as they were advertised to be.[17]

Representatives of the electrical industry saw domestic appliance use in an idealized way. Although they stressed values of efficiency and economy in their sketches of saved cost and labor, "the coziness and comfort of preparing coffee on the table," as one SCE agent's letter to housewives put it in 1911, was the thing housewives really would appreciate. Moreover, the men appreciated it. After all, home was their retreat from the "technological order," and even if they were imposing such an order on women in the domestic sphere, they had good psychological reasons not to acknowledge it. For his part, Edgar Wilcox assumed that some housewives preferred cooking entire meals at the table rather than in the kitchen. After describing "table cooking outfits" with electric-plate stoves using hollowware utensils and accompanied by electric percolators, tea samovars, chafing dishes, frying pans, and griddles, he concluded: "For the hostess who does her own cooking at the table, cooking outfits are ideal . . .[not to mention] an ornament to any sideboard or table." In 1915 S. M. Kennedy concluded that housewives were using lamp-socket percolators, toasters, chafing dishes, and grills to cook breakfast and lunch at the table because economic hard times meant that housewives had to get along without help. Thus, tabletop cooking made things easier by "saving many steps to and from the kitchen."[18]

In 1915 West Coast electric power companies began earnest marketing of electric ranges and other large appliances. They advertised "cooking by wire" as efficient, without guesswork, clean, and odor free. As SCE put it in a letter to housewives: "A party dress is perfectly safe and the work is SO much easier. Meals are cooked by the turn of a switch and toil and soil are no more." And Wilcox's advice to salesmen about selling ranges reveals even more about attitudes toward women and machinery: "Being clean, safe and labor-saving, the extent of the improvement [in cooking] brought about by the electric range is almost unbelievable." It was so simple, said Wilcox, that "the operator has only to watch the clock while the food is cooking. . . . Any housewife, of even moderate intelligence, should be able to master the essential features of the operation of an electric range in a short time."[19]

Wilcox's tone revealed a thinly veiled view that although women might not be technologically competent, they were compliant, pliable, and willing to learn to use any new domestic appliance. Thus, he urged cultivating electric cooking early among women, placing appliances in domestic-science classrooms and purchasing young girls the "Hughes Junior Range (for Early Training of Housewife)." In the end, he argued, "the housewife who does her own cooking is the most desirable user of an electric range. She will be, as a rule, thoroughly alive to its advantages, will practice the many little economies that are possible, and will generally become a 'booster' for electric cooking." Salesmen should therefore "give the user early and painstaking attention." "It must be remembered," he said, "that the manipulation of an electric range is entirely new to the average housewife." Any troubles not "rectified or explained away . . . will become magnified as time passes, and the housewife may finally become seriously prejudiced." Although Wilcox did not expect the housewife to understand the technology, he fully anticipated that she would accept explanations given her. She was easy prey for salesmen to handle compared with "a professional cook." "This type of individual," he said, "is frequently a difficult person to handle. He seldom favors anything new. He is prone to form intense prejudices; and will often refuse to make an intelligent investigation of new apparatus, especially when he has not been previously consulted." He was an enemy. Salesmen could not easily manipulate the male range operators nor "explain away" problems when dealing with them.[20]

An electric range, unlike an individual lamp-socket appliance, was a primary machine in a complex system of food preparation. Wilcox encouraged range salesmen to "know something about the use to which [the range] is to be applied" and offered them a bit of information on how food should be cooked and different ways to cook it. Similarly, industry sales managers on the West Coast sought "firsthand knowledge" about using electric ranges by placing them in their own homes. Nevertheless, men had a difficult time introducing the electric range into the intricacies of a food-preparation system that they hardly understood. Some of them, like Wilcox, turned to marketing commercial and industrial electrical technology. In 1928, when the McGraw-Hill Book Company brought out an updated edition of Wilcox's book, it was addressed to contractors, not salesmen. He gave scant attention to domestic appliances and focused instead on commercial and industrial tools—real technology for real men. In a sense he had devalued technology associated with women, but perhaps there is more. When ranges were very new Wilcox felt that women greatly appreciated them. In anticipating their excitement over a new range, he was like a young husband eager to bring home a gift for his bride. The same delight hardly colored his view about selling ranges to professional cooks. But the range, he discovered, was a complex gift. Ultimately Wilcox found it more comfortable to interact with men.[21]

Inclusion of Women in Marketing

About the time utility managers began to perceive the load and revenue benefits of electric ranges, nearly all West Coast power companies employed some women in their marketing efforts. Although managers probably did not imagine that their industry might unsettle home life by

4.3. Mrs. Colby, in white, and her assistant, Southern California Edison's motherly appliance demonstrators in 1912. Courtesy Southern California Edison Company.

deluging housewives with appliances, it nevertheless interfered in a sphere in which female elders traditionally taught their daughters how to run things. Thus it was propitious that women joined the sales arm of the industry in the early 1900s to demonstrate the widening variety of domestic appliances. Some women demonstrators appeared to have been ordered up from Hollywood central casting as mother substitutes. If electric ranges were to be successfully placed in homes, women in the industry acting as surrogate female elders would be helpful. Certainly salesmen needed all the help they could get to successfully introduce the technology of electric power into the complex system of domestic food preparation. As Wilcox had discovered early on, one had to do more than just place the range in the kitchen (fig. 4.3).[22]

Over the years husbands rather than wives commonly had visited showrooms and purchased appliances for the household. "For several years," reported an anonymous writer for San Francisco's *Journal of Electricity* in 1923, "the entire electrical industry, whenever it had something to sell, made its approach to the men."[23] During the 1910s, however, a role for women in marketing electric appliances began to take shape. Women appeared using appliances in advertising photographs and regularly demonstrated ranges and other domestic tools at power company offices. The New York–based trade journal *Electric World* reported in 1916 that most central stations also sent "an expert in cooking on electric ranges (usually called a 'home economist') to call on the purchaser a few days after the range is installed." Men in the industry gradually began to acknowledge that women should play a larger role in appliance marketing. They seemed to recognize, as Carolyn Goldstein has written, that "fitting electricity to the home required fitting women to the electrical business."[24]

The anonymous writer for the *Journal of Electricity* made the observation on marketing without women before the 1920s in an article describing SCE's new range campaign and then went on to suggest that the firm's managers now saw the incongruity of their marketing method: "[Approaching men] was all very well as regards motors and industrial apparatus but did not work out in the case of domestic appliances and resulted in the impedance to the sale of this type of equipment. No man would permit his wife to tell him what kind of equipment he should have in his office or factory, what kind of machinery to buy and how to run it. Yet men for some years were looked to to decide what kind of machinery and equipment the housewives should have for running their businesses. An anomaly, to be sure; a paradox, indeed." The men who formulated SCE's plan, the writer continued, "wisely decided to give due recognition to the person who would actually use or be most concerned in the use of the equipment, namely, the housewife, and all the campaign material was directed at her." To be sure, this was a major change in marketing strategy, and the drive's advertising copy suggests that women participated in its creation. But the material way in which the company actually carried out face-to-face contact with customers made it plain that the real selling of appliances was still gendered. At local demonstrations, important events in the marketing campaign, "trained domestic science experts" demonstrated the goods, but "salesmen" were on hand "to explain operation of equipment and rates and to secure the names and addresses of those whose interest warranted follow-up."[25]

Clotilde Grunsky, one of three women members of the American

Institute of Electrical Engineers in 1922, may well have been the anonymous writer for the *Journal of Electricity.* After graduating Phi Beta Kappa and receiving the faculty award as the most distinguished graduate of the University of California in 1914, Grunsky did "vocational and welfare work among women" in the San Francisco Bay Area and lectured at Mills College on "vocational problems." The daughter of a noted West Coast civil engineer, she was attracted to her father's profession, but there is no evidence to suggest that she ever sought to practice electrical engineering. If she had, she certainly would have faced enormous obstacles, not the least of which would have been, as sociologist Judy Wajcman has suggested, forsaking her femininity in order to work with the boys. By 1917 she was the associate editor of the *Journal of Electricity,* which soon boasted that she was "recognized nationally as perhaps the best woman writer on technical subjects pertinent to the electrical industry."[26]

Grunsky and other women in the industry championed both the housewife's electrical needs and women's working within electric utility organizations. In 1921, when the National Electric Light Association established a Women's Committee on Public Utility Information within its Public Relations Section, Grunsky was a member, and she led the women in California's electric industry as chairperson of the Woman's Public Information Committee of the Pacific Coast Electrical Association. The committee sought ways to best get the electric message to housewives, urging use of the radio, organizing an industry information service, and encouraging development of power company home economics laboratories. In 1925 the committee surveyed twelve California and three Arizona power companies about women's roles in the industry. Most of them employed from one to seven women demonstrators, five maintained demonstration and exhibition kitchens, and six or seven ran cooking schools, alone or with local newspapers or manufacturers. Their women employees helped develop newspaper advertising and company-sponsored newsletters, such as *P.G.&E. Progress,* in which one page of each issue was devoted to "Help for the Homemaker." Finally, they reached housewives informally by joining local women's clubs (with dues paid by their companies), and they served as ambassadors for home electricity at county fairs and other community events.[27]

Women's committees helped both to professionalize the role of women within the industry and to involve them directly in sending the electric message to all women. Their influence was widespread and could be seen readily in displays such as the one women conceived for the San Joaquin Light and Power Company at a Fresno district fair during the 1920s. Integrated as an essential part of the exhibit was a free day nursery, a restroom, a children's playground, and an emergency hospital. Female attendants watched children in the playground, nurses cared for infants in the nursery, and there was always a physician present. Electricity ran a refrigerator to keep milk fresh and heated milk in a warmer. Regular meals also were provided. On a year-round basis the firm maintained another exhibit at the Fresno free market, which was set up as a "model wash room and sewing room" staffed by women demonstrators.[28]

Women brought a different perspective and sensitivity to electric appliance sales, but men still dominated the process. In 1923 SCE used "a corps of specially trained men" to present public electrical demonstrations and illustrated slide lectures. Through the 1930s men demonstrated new appliances and even trained women home economists to use them.[29] Women may have been brought into the marketing operation, but men remained firmly in control of how housewives received the electric message.

Conclusion

The harnessing of electricity spawned a prodigious procession of new domestic tools during the first decades of the twentieth century. Consciously and unconsciously, gender stood as an important ingredient in the marketing of electric appliances. The fact that men marketed tools designed to better carry out tasks residing within the sphere of women made the issue explicit, and it was further highlighted by power company engineers' and salesmen's peddling lamp-socket devices as convenient appliances for housewives when in truth they placed their own firms' convenience first. More subtly, the language and approach men selected in speaking to and about women consumers, their misconceptions about women in the home, their mission to train housewives in technology, and their seeing women as prey while they saw men as adversaries in the marketing process betrayed deep-seated beliefs and had the potential to cripple utility marketing strategies. Thus, by including women in marketing operations, utilities sought to explicitly appropriate gender to the selling of domestic electric technology.

Finally, examining the role of gender in the marketing of electric appliances perhaps helps clarify how differently men and women relate to technology. For men, domestic technology meant conquering household inefficiency. Control, power, mastery—that was the stuff of real men. For women, domestic technology meant household choices. Negotiation, rearrangement, and tinkering were the stuff of real women. Men may unconsciously have tapped into women's real experiences with the tedium of housework and offered women greater flexibility in carrying out some household tasks through electric appliances, which no doubt helped them in their marketing efforts, but the meaning and value of domestic technology continued to remain different for men and women.

Notes

1. David E. Nye, *Electrifying America: Social Meanings of a New Technology, 1880–1940* (Cambridge: MIT Press, 1990), is essential for understanding the broad societal impact of electricity. On the introduction of electric tools into the household see Ruth Schwartz Cowan, *More Work for Mother: The Ironies of Household Technology from the Open Hearth to the Microwave* (New York: Basic Books, 1983); and Susan Strasser, *Never Done: A History of American Housework* (New York: Pantheon, 1982). Mark Rose, *Cities of Light and Heat: Domesticating Gas and Electricity in Urban America* (University Park: Pennsylvania State Univ. Press, 1995), discusses introducing electricity into the home in Kansas City and Denver; and Carolyn M. Goldstein, "From Service to Sales: Home Economics in Light and Power, 1920–1940," *Technology and Culture* 38 (Jan. 1997): 121–52, examines the role of women home economists as mediators between the electric power industry and women consumers. Judy Wajcman, *Feminism Confronts Technology* (University Park: Pennsylvania State Univ. Press, 1991), provides crucial insight into the issue of gender and technology, and I frequently draw from her work in this chapter.

2. Thomas P. Hughes, *Networks of Power: Electrification in Western Society, 1880–1930* (Baltimore: Johns Hopkins Univ. Press, 1983), 218–19.

3. On California's hydroelectric industry see James C. Williams, *Energy and the Making of Modern California* (Akron OH: Univ. of Akron Press, 1997), ch. 8.

4. S. M. Kennedy, "Sale of Electrical Appliances for Regular Lamp Circuits and

Their Effect on Load and Income," *Electrical World* 60 (7 Dec. 1912): 1210. The population of the nine-county southern California area grew from 308,588 in 1900 to 1,354,081 in 1920. A general history of SCE is William A. Myers, *Iron Men and Copper Wires: A Centennial History of the Southern California Edison Company* (Glendale CA: Trans-Anglo, 1983).

5. Kennedy, "Sale of Electrical Appliances for Regular Lamp Circuits," 1210; S. M. Kennedy, "Southern California Edison Company's Method of Disposing of Electric Appliances," *Electrical World* 62 (22 Nov. 1913): 1062–63.

6. Kennedy, "Sale of Electrical Appliances for Regular Lamp Circuits," 1210; A. W. Childs, "Building an Appliance Load," *Journal of Electricity* 49 (1 Dec. 1922): 404–5. On lamp-socket appliances see Fred E. H. Schroeder, "More 'Small Things Not Forgotten': Domestic Electrical Plugs and Receptacles, 1881–1931," *Technology and Culture* 27 (July 1986): 530–33.

7. Kennedy, "Sale of Electrical Appliances for Regular Lamp Circuits," 1210–11; S. M. Kennedy, "Selling Lamp-Socket Appliances," *Electrical World* 65 (29 May 1915): 1412–14. The population served by SCE was about five hundred thousand in 1915. General Electric was so impressed by one photograph of a pile of two thousand flatirons taken in trade by SCE that it circulated it widely within the industry (see Nye, *Electrifying America,* 264. The photograph also appears in Kennedy's article "Southern California Edison Company's Method of Disposing of Electric Appliances," 1063; and a photograph of the same scene from a different angle appears in Childs's article "Building an Appliance Load," 405). In 1908 Chicago's Commonwealth Edison successfully adopted SCE's strategy of lending electric irons (see Richard F. Hirsh, *Technology and Transformation in the American Electric Utility Industry* [Cambridge: Cambridge Univ. Press, 1989], 212 n. 18).

8. Ruth Schwartz Cowan, "The Consumption Junction: A Proposal for Research Strategies in the Sociology of Technology," in *The Social Construction of Technological Systems: New Directions in the Sociology and History of Technology,* ed. Wiebe E. Bijker, Thomas P. Hughes, and Trevor J. Pinch (Cambridge: MIT Press, 1987), 261–80. Rose, *Cities of Light and Heat,* 81–86, discusses gender and marketing in terms of the Denver Gas and Electric Company.

9. The identification of the male with technology and progress has been discussed widely. See esp. Judith A. McGaw, "No Passive Victims, No Separate Spheres: A Feminist Perspective on Technology's History," in *In Context: History and the History of Technology: Essays in Honor of Melvin Kranzberg,* ed. Stephen H. Cutcliffe and Robert C. Post (Bethlehem PA: Lehigh Univ. Press, 1989), 172–91; John M. Staudenmaier, *Technology's Storytellers: Reweaving the Human Fabric* (Cambridge: MIT Press, 1985), a work that has at its core the myth of progress and is very cognizant of gender; Joan Rothschild, ed., *Machina Ex Dea: Feminist Perspectives on*

Technology (New York: Pergamon, 1983), xviii–xxii; and Wajcman, *Feminism Confronts Technology,* 1–25. Finally, on the subject of technological progress see George Basalla, *The Evolution of Technology* (New York: Cambridge Univ. Press, 1988), esp. 210–18.

10. Frank Baum, "The Effect of Hydro-Electric Power Transmission upon Economic and Social Conditions, with Special Reference to the U.S. of America," in *Transactions of the International Engineering Congress, 1915,* vol. 7 (San Francisco, 1916), 247; Mr. Junkersfield, of Chicago, in "Discussion" following S. E. Doane, "Electric Light—A Factor in Civilization," *Journal of the Western Society of Engineers* 20 (Jan. 1915): 17. On the virtues ascribed to electricity and its centrality to modernity see Nye, *Electrifying America;* James W. Carey and John J. Quirk, "The Mythos of the Electronic Revolution," *American Scholar* 39 (spring 1970): 219–41 and 40 (summer 1970): 395–424; David Nasaw, "Cities of Light, Landscapes of Pleasure," in *The Landscape of Modernity: Essays on New York City, 1900–1940,* ed. David Ward and Oliver Zunz (New York: Russell Sage Foundation, 1992), 273–86; and contemporary sources such as R. B. Owens, "Electricity as a Factor in Modern Development," *Cassier's Magazine* 16 (June 1899): 212. See also George Basalla, "Some Persistent Energy Myths," in *Energy and Transport: Historical Perspectives on Policy Issues,* ed. George H. Daniels and Mark H. Rose (Beverly Hills: Sage, 1982), 27–38.

11. Quoted in "Popular Reflections," *Journal of Electricity, Power, and Gas* 1 (July 1895): 27.

12. Rose, *Cities of Light and Heat,* 78–79.

13. Sherry Turkle, *The Second Self: Computers and the Human Spirit* (London: Granada, 1984), cited in Wajcman, *Feminism Confronts Technology,* 156. Richard Hirsh provides a good example of masculine "hard mastery" over technology in *Technology and Transformation,* his study of the evolution of steam plants in the electric power industry.

14. Kennedy, "Sale of Electrical Appliances for Regular Lamp Circuits," 1212.

15. Edgar A. Wilcox, *Electric Heating* (San Francisco: Technical Publishing, 1916). *The Journal of Electricity, Power, and Gas* summarized Wilcox's book between Nov. 1916 and April 1917, ensuring it a wide regional audience.

16. Wilcox, *Electric Heating,* 8. The Hotpoint Electrical Company, started in Ontario, California, had such success with its iron that General Electric acquired the firm and its line of products during the 1920s (see Nye, *Electrifying America,* 262, 265).

17. Schroeder, "More 'Small Things Not Forgotten,'" 531.

18. Advertising letter in Kennedy, "Sale of Electrical Appliances for Regular Lamp Circuits," 1210; Wilcox, *Electric Heating,* 12–13; Kennedy, "Selling Lamp-Socket Appliances," 1414. Ruth Schwartz Cowan makes the point that the household was seen as "a place where men could retreat from the technological order" (60) in her

Bicentennial article "From Virginia Dare to Virginia Slims: Women and Technology in American Life," *Technology and Culture* 20 (Jan. 1979): 51–63; see also Wajcman, *Feminism Confronts Technology,* 89–90.

19. S. M. Kennedy, "Cooking by Wire in Southern California," *Electrical World* 67 (20 May 1916): 1169; Wilcox, *Electric Heating,* 26–28.

20. Wilcox, *Electric Heating,* 31–34.

21. Ibid., 38–41; "Establishing the Electric Range in the Far West," *Electrical World* 67 (22 Apr. 1916): 926; E. A. Wilcox, *Electric Heating,* rev. ed. (New York: McGraw-Hill Book Company, 1928). Wajcman, *Feminism Confronts Technology,* 24, 37–39, speaks to the issue of masculinity and technology.

22. Glenna Matthews, *"Just a Housewife": The Rise and Fall of Domesticity in America* (New York: Oxford Univ. Press, 1987), ch. 6, discusses the relationship between women in the domestic-science movement and tradition. Initially tradition and progressivism collided, but over time the home economist assumed the role of female elder, more than effectively replacing the wisdom of "mother" (156). See also "Establishing the Electric Range in the Far West," 926.

23. "Selling Electric Ranges in Southern California," *Journal of Electricity* 51 (1 Oct. 1923): 251. On men as purchasers of electric technology see Louie Carlat, "Gender and the Domestication of the Radio, 1922–1935" (paper delivered at the annual meeting of the Society for the History of Technology, Washington DC, 17 Oct. 1993), which points out that "while the person twisting the dials at home might be female, the typical purchaser in the twenties was still male" (3). A revised version of Carlat's essay is published as chapter 5 in this volume.

24. Goldstein, "From Service to Sales," 128; "Establishing the Electric Range in the Far West," 926.

25. "Selling Electric Ranges in Southern California," 251–53. Goldstein, "From Service to Sales," 150, underscores that, as "home service departments" established within utilities during the early 1920s expanded, "home economists became part of a gendered division of labor in which women gave instruction about appliances and men sold them."

26. Information about Grunsky appears in the personal columns of the *Journal of Electricity* 41 (15 Oct. 1918): 347; the *Journal of Electricity and Western Industry* 48 (15 Feb. 1922): 173; and the University of California yearbook, the *Blue and Gold,* vols. 41 (1915): 260 and 42 (1916): 41. To my knowledge, Grunsky never published under her own by-line in the *Journal,* but she handled almost everything dealing with marketing and public relations. She is cited as the chair of the Women's Public Information Committee of the Pacific Coast Electrical Association, whose 1925 report the *Journal* published ("Public Relations from the Woman's Standpoint," *Journal of Electricity* 54 [1 June 1925]: 530–34). Carolyn Goldstein, "From Service to

Sales," 131 n. 27, cites Grunsky's article "Always with a Smile!" *Electrical Merchandising* 31 (Mar. 1924): 4148–79, 4172. Finally, see Wajcman, *Feminism Confronts Technology*, 19.

27. "Public Relations from the Woman's Standpoint," 530–34.

28. Ibid., 532; "Space in Free Market Utilized for Display Booth," *Journal of Electricity and Western Industry* 47 (1 Oct. 1921): 274.

29. Southern California Edison Company, *Annual Report for the Year 1923* (Los Angeles, 1924), 22. Home economists are not mentioned in SCE's annual reports before the report for 1930, and then they are named as working with "highly trained specialists" (*Annual Report for the Year 1930* [Los Angeles, 1931], 24).

Five

"A Cleanser for the Mind"

Marketing Radio Receivers for the American Home, 1922–1932

Louis Carlat

In 1922 a junior executive of the Radio Corporation of America made a prediction to a conference of electrical equipment dealers. David Sarnoff, who later became chairman of RCA, promised the dealers that radio, then only a popular fad, could become a household appliance unlike any other. Likening it to a more familiar technology, indoor plumbing, he forecast that it would grow into an equally common part of domestic life. "I believe the day will come," he declared, "when a man will no more think of failing to equip his home with a radio receiving device than [he would] today think of failing to equip [it] with a bathtub. One constitutes a cleanser for the body and the other for the mind."[1]

Unlike other supposed forecasts that Sarnoff made after the fact, this one really was prescient. With help from Sarnoff and RCA, radio did become an appliance even more common than indoor plumbing.[2] Between the early twenties and the mid-thirties it was transformed from a male-dominated hobby of the garage or attic into a furnishing for respectable family spaces. Changes in radio technology altered old ways of listening; with better reception, new receivers offered novel pleasures and invited inexperienced users. Although receivers that were easier to operate made new listening habits possible, persuading consumers to adopt unfamiliar behaviors required both concerted advertising efforts and new cultural attitudes toward radio technology. After RCA and similar firms marketed receivers so successfully during the twenties, women no longer needed to suffer as radio widows, which one woman defined as "being married without a husband—or at least, a radio husband—which amounts to the same thing." Instead of being excluded while their husbands soldered wires and fiddled with erratic reception, women learned to participate in the pleasures of domesticated listening.[3]

The transition from male toy to a component of domestic space required recasting radio hardware as a feminine object, and listening as a feminine activity. Advertising's assignment of gender categories to the various capabilities of receivers illustrates competing definitions of what receivers could become, and for whom. Using radio to explore aurally distant cities was invariably portrayed as male, while its use to acculturate husbands and children was invariably seen as female. Which gender advertising targeted depended on the model advertised.

Functionally receivers remained essentially unchanged between 1922 and 1935. A model from either year could detect radio signals and amplify them into intelligible sound. In social terms, however, these receivers were entirely different devices. The earlier one was promoted for men and boys to extend human hearing, conquer distance, and "listen in" on unseen cities. The later models were advertised for their putative ability to maintain the boundaries between domestic and outside spaces. In the face of apparent threats to middle-class family life radio might encourage women to stay safely inside the confines of their homes and traditional roles.

Makers and promoters of radio sets made significant distinctions between men and women. Receivers were redesigned and mass-manufactured specifically to make them more accessible to women, but advertising for them was aimed almost exclusively at men. These strategies reflect concerns about changes in leisure, domesticity, and gender roles and uncertainty about radio's place in the home and in American culture.

This chapter examines radio's transition from a device built by amateurs into a mass-produced cultural instrument for middle- and upper-class consumers. How did manufacturers tailor their products to new markets, and how did advertisers identify and persuade possible consumers? Radio was hailed in the popular press as an inherently universal and democratic medium, but the commercial diffusion of receiving devices was highly selective, guided by conceptions of class and especially gender.

We know from the work of historians Susan Douglas and Susan Smulyan that operators in attics, in garages, and clandestinely under beds "listened in," as the contemporary phrase had it, to the ether waves. Both in fact and in advertising representations these operators were predominantly male. Amateurs formed an invisible network of individuals swapping stories and information about their lives and their towns. Monitoring or joining these conversations provided entertainment for those who could not or would not participate in crowded public amusements or those who simply hungered for direct contact with the world beyond the horizon. Equipment advertisers and hobby magazines encouraged home operators, nicknamed "DXers," to keep track of cities they heard and to compete with friends for the greatest distances attained.[4]

The ability to manipulate a receiver to bring in signals from a faraway transmitter was respected within the culture of amateur wireless. It might even lead to greater glory. One maritime operator, Jack Binns, won instant fame for orchestrating the rescue of passengers of the *Republic*, a crowded luxury liner that sank near Nantucket in 1909. In the foreword to a novel about the adventures of a group of young radio enthusiasts, Binns admonished, "Radio is still a young science, and some of the most remarkable advances in it have been contributed by amateurs—that is, by boy experimenters. It is never too late to start in the fascinating game, and the reward for the successful experimenter is rich both in honor and recompense. . . . Don't be discouraged because Edison came before you. There is still plenty of opportunity for you to become a new Edison, and no science offers the possibilities in this respect as does radio communication." The art of "listening in" had less to do with absorbing content than with manual dexterity, technical mastery, and receiving from as far away as possible.[5]

This situation began to change rapidly around 1922, when the popular press and large corporations began to take notice of the amateur phenomenon. Companies that owned important radio patents abandoned their original plans to develop wireless as private point-to-point communication from a transmitter to a single receiver. Manufacturers such as RCA, Westinghouse, and General Electric recognized the popularity of more generalized "listening in." Fueled by the press's sudden interest in a phenomenon it had previously ignored, wireless enthusiasm grew into a popular craze in the early twenties. Trying to satisfy the public's curiosity, publishers launched new magazines about building and selling wireless. No fewer than ten new periodicals appeared between 1920 and 1925, with seven going to press for the first time in 1922, at the height of the craze. Builders, wholesalers, and retailers of radio equipment advertised heavily in these new publications.[6] Although these magazines expected women's involvement with radio technology to be primarily in the manufacturing process (e.g., as testers of vacuum tubes), they occasionally reported on women and girls who had taken it upon themselves to learn (and teach others) how to operate receivers. *Radio World* even began a column entitled "The Radio Woman." Of course, the effort to broaden radio's constituency sometimes met with resistance from more conventional enthusiasts. One young woman recounted that when her brother failed to ground his set properly, she studied *Radio World* and "memorized some of the technical phrases . . . and sprang them on Jimmy, who nearly had a Dutch fit at my 'familiarity' with his pet hobby." Brother Jimmy also had to admit that the writer's idea of locating the receiver next to the phonograph, his family's traditional center of musical culture, was a good one.[7]

Trying to capitalize on the boom market, manufacturers moved quickly to develop broadcasting stations as a means of stimulating sales of receiver equipment. This irreversibly shifted radio's technical and economic development from point-to-point communication to broadcasting, in which a signal is transmitted in all directions and anyone within range can pick it up. Popular speech quickly reflected the change. What had been known as *wireless,* a shorthand for *wireless telephony* or *telegraphy,* by the early twenties was known as *radiotelephony,* or more commonly, *radio.*[8] The *New York Times* adopted the new term in 1924.

By the late twenties a new group of auditors had overwhelmed the ranks of amateur wireless operators. These were consumers of factory-produced receivers, people who did not necessarily understand how radio worked but who wanted to enjoy the growing number of professional broadcasts in the comfort of their living rooms.[9] The technology for receiving electrical signals through the air came to connote different settings and users. At one moment *wireless* meant one thing, but a decade later *radio* commonly was understood as a very different device, used for entirely new reasons.

The historiography of American radio retains a curious blind spot, perhaps the result of its emphasis on and indebtedness to the broadcast industry. Most of the historical literature focuses on the technology, the economics, or the regulation of the production of broadcasting. But what about its reception? The importance of having a mass-produced radio receiver in the home in the first place has been largely neglected.[10] Transmitters without receivers are useless; broadcasters without listeners are merely talking to themselves. Companies such as Westinghouse Electric and American Telephone and Telegraph (AT&T), which established early broadcast stations in the hopes of selling receivers and advertising time, could transmit signals for as long as they were willing to pay expenses, but who would listen, and how?[11]

Standard works such as Eric Barnouw's three-volume history of broadcasting have treated the growth of the radio audience and the receiver market as the inevitable result of consumer demand for improved receivers. In her otherwise excellent account of the invention of network broadcasting historian Susan Smulyan suggests a different kind of determinism, in which "technological improvements in receiver design worked against long-distance listening because ready-made radio receivers, with improved reception quality, created a new type of radio fan," one interested in program content. Susan Douglas's *Inventing American Broadcasting* explores the role of the press and broadcasters themselves in stimulating consumer enthusiasm but, largely because it ends with 1922, does not consider the effects of direct advertising of radio equipment, nor does it reflect the differentiation of this growing new market along class and gender lines. Recent studies of listening behavior, as welcome as these are, assume the uniform availability of radio equipment. One exception is Lizabeth Cohen's richly detailed study of industrial workers in Chicago between the world wars. Cohen notes that blue-collar families generally could not afford factory sets, so neighbors often pooled money to buy one for an entire block to share. They may also have kept alive the hobbyist's knowledge and skill, building their own sets from parts and scrap. Yet despite their great numbers, because workers like these often could not afford manufactured radios, at first receiver makers did not consider them a significant market.[12]

Americans of means, however, enjoyed a different situation. During the twenties they were learning to purchase and use household machines, especially electrical equipment, although doing so did not come naturally. The public had to be educated to desire and evaluate machines for convenience and home entertainment. In short, consumers had to be created, whether for vacuum cleaners, automobiles, or radios. This required concerted promotional strategies such as an annual Radio World's Fair at New York's Madison Square Garden, at which more than three hundred manufacturers introduced their latest models to the public. Expositions such as this were important industry events and attracted considerable press coverage. The annual model change, a practice usually associated with the auto industry, was also employed by radio manufacturers, who introduced new or revised product lines each year through the twenties. Doing so enabled retailers to encourage consumers to upgrade equipment. All the major radio shows and most new equipment debuts occurred in late summer or early fall to accommodate holiday shopping and the increased popularity of radio listening on clear, cold nights.[13]

Radio ownership in the twenties was sharply divided along class lines. Advertisements showed radios in elegant, even opulent surroundings, emphasizing that this was a cultural luxury or status item. This also reflected the fact that factory-built receivers frequently cost more than a hundred dollars, putting them out of the reach of many families. Despite manufacturers' fantasies of a mass market, fully assembled receiver sets remained instruments of middle- and upper-class leisure during the twenties. This market was sizable, to be sure, but hardly large enough to make radio the universal medium its backers hoped it would become. Not surprisingly given the distribution of wealth in the United States in 1930, an overwhelming majority of receivers in that year were in the homes of white Americans, particularly the native-born living in large cities.[14]

Perhaps because they concentrated their efforts on the most affluent consumers, receiver advertisers regularly aimed their copy a step or two above their audiences. Readers of mass-circulation middle-class magazines like *McCall's* saw illustrations of elegantly dressed men and women enjoying radio while relaxing in spacious homes. Depicted this way, radio was for the most upwardly mobile, modern families. In a 1929 series of ads for Eveready the N. W. Ayer Advertising Agency stated that Eveready constructed its receiver for "the man who lives for the song of a golf-ball in full flight . . . [and] the woman who drives her own sports-car." The Eveready ad promised both cultural refinement and amusement for those who had, or could hope to have, sufficient money and leisure "to enjoy the better things of contemporary life."[15]

Despite having been commercialized for mass consumption, the old wireless remained the functional equivalent of the new radio. The most humble 1920 homemade crystal receiver set was designed to perform the same basic functions that an Ultradyne, Atwater Kent, Philco, or any other brand flooding the market could handle. Any of these devices could detect radio signals and transform them to intelligible sounds of music or speech more or less amplified. Of course, internal circuits and methods of tuning differed, but the end result, audible messages carried through the ether, was the same. And yet, as social technologies integrated into everyday life, they differed greatly. These comparable devices—the wireless and the radio—were used in different contexts, for different purposes, by different operators.

Advertisements highlighted the contrast between old and new. One manufacturer sought middle- and upper-class consumers ready for "a new conception of radio" in 1925. The Phenix Radio Corporation boasted that its Model L-3 Ultradyne receiver was a "new kind of radio-musical instrument" in which the best scientific knowledge had been distilled for easy use at home. Below a drawing of an meticulously dressed man and woman seated before the L-3 in a spacious, modern living room, the ad copy promised that "the Ultradyne Receiver is worthy of the place of honor in the most luxurious homes." The radio merited such high status because the cumbersome, inelegant radio device of the past had finally been simplified and nested in a mahogany cabinet suitable for domestic display. With this modern receiver, the advertisement exulted, art had triumphed over "mere mechanics."[16]

Whatever the actual merits of the new model, its promotional campaign epitomized the transformation of wireless from a workshop curiosity, what Ultradyne called "a laboratory machine," to an important household furnishing and a focus of family entertainment. The hobbyist's wireless had become a piece of furniture off limits to tinkerers. Ultradyne enforced this change with a promise to consumers. Each of its receivers left the factory with a monogrammed seal placed on its assembly lock bolts. "To protect the public" ignorant of the delicate circuits within, Ultradyne warranted each receiver against defects "so long as the seals remain unbroken." To tinker would be to transgress.[17]

The newer receivers were designed to make tinkering unnecessary, if not impossible. A *New York Times* writer observing this trend in 1928 praised that year's models, in which "tubes, transformers, condensers and wiring, are concealed in metal compartments, making radio more efficient and less likely to develop trouble." Of course if something did go awry, the owner would have to call upon a professional technician rather than an electrically minded neighbor, brother, or son. Retailers took this opportunity to enhance their reputations through the quality of their repair service. Retailing magazines urged aspiring dealers to consider service an important part of their business because they would be selling to people incapable of making their own repairs, and good service after the sale would ensure repeat business.[18]

The advantage of making home adjustments unnecessary lay in its appeal to a new kind of purchaser and user. The parts business had always depended on the hobbyist who built and maintained his own sets. However, astute manufacturers recognized that their traditional customers formed but a fraction of the potential market. They began to design receivers that would appeal to a new constituency: women. Aided by advertising, they changed the technology and its image to attract a wider and more feminized market, one unaccustomed to electro-mechanical tinkering. They reinvented radio as a factory-produced artifact whose operation required less technical skill on the part of users. Builders also created new social uses for their wares. Their ads began to deemphasize the thrills of distance and exploration and recast the technology's image as part of more domestic scenes. Illustrated advertisements in trade and sales magazines shifted from depicting a boy or group of males to showing families or mixed clusters of men and women. The device was portrayed as having uses other than putatively masculine competition and exploration. As early as 1924 it was even possible to associate radio with feminine stereotypes, as did a cartoon of two women engrossed in conversation above the caption, "The first broadcasting station." Both gossip and radio, this ad suggested, were forms of communication that under the best circumstances could be genuinely informative and useful.[19]

Getting radio in the hands of new consumers required making the devices easier to use. The old crystal sets had serious limitations of detection (finding and holding a signal) and amplification. They were fickle and notoriously difficult to tune, and they produced weak sounds audible only through individual headsets. The replacement of crystals with vacuum tubes by the early twenties increased the sensitivity of detection circuits, that is, their ability to hold a signal at a given frequency. Regenerative circuits, in which part of the vacuum tube's electrical output is returned to it, multiply the signal's power and increase the volume of sound produced. With regeneration, amplification was high enough that users no longer needed to wear individual headsets connected to the receiver by hollow tubes through which the sound traveled; instead, acoustic horns like those used on phonographs could project the sound into an entire room.

In addition to raising volume, regeneration also increases a circuit's ability to detect faint signals (sensitivity) and to distinguish them from other transmissions at similar wavelengths (selectivity). However, this requires carefully controlling both the degree of regeneration and the voltage supplied to the filament and plate, separate components within each tube. Regulating the current required a separate control for each tube and was particularly tricky with battery-powered receivers. Too much regeneration or a change in the balance between filament and plate voltages could cause a circuit to oscillate, producing the distinctive shrieks of feedback. Worse, these oscillations could interfere with other receivers in the vicinity. Controlling regeneration and voltage to as many as five or six tubes for each receiver took skill and practice. As *Radio World* cautioned, using a receiver properly "is not just the vain twisting of a control dial until something happens; it is far more than that—it is a systematic, scientific manipulation of variable conditions so that the desired results are obtained. The man who just turns dials in a childish and unknowing fashion waiting for the 'magic box' to spring a 'hocus pocus' trick is not only cutting himself off from a lot of entertainment but is also seriously hurting his friend nearby." Advice in this and other hobby magazines on manipulating regeneration did not eliminate occasional complaints from readers about radio's squawking irritations.[20]

Tuning the early vacuum tube receiver was a problem the operator had to solve anew each time he selected a different station. This was usually accomplished by physically rotating the metal plates of a capacitor to vary the circuit's capacity, allowing it to respond to different wavelengths and effectively tuning it to the frequencies of various transmitters. This would be a simple task for a single circuit. Most receivers, however, had at least two such independent circuits to increase their sensitivity, and these had to be tuned together. The most direct way to do this was by means of an independent control for rotating each capacitor.[21] To tune each circuit to the same wavelength, an operator again had to develop both a "feel" for the mechanism and the patience to make minute adjustments to each of several controls. Because of mechanical and electrical variability in these controls, tuning to a given frequency might require a different configuration of the dials each time. Only as circuit responses were made more uniform at mid-decade did designers dare to use dials with calibrations already marked on them.

Manufacturers recognized the vagaries of tuning multiple-circuit receivers as the major obstacle to increasing radio use. A receiver with a single tuning mechanism would enjoy greater popularity, but first significant electrical and mechanical difficulties had to be overcome. According to historian Arthur Harrison, the search for single-control tuning produced well over a hundred proposed solutions in the 1920s. Obviously, this flurry of creativity ended well for manufacturers. The "gang tuner," which linked capacitors by a variety of ingenious means, reduced tuning to the relatively easy manipulation of two dials. In 1925 the Mohawk Electric Corporation introduced a true single-tuning radio, operated by one control. By 1928, after problems with the Mohawk's styling and construction had been resolved, single-tuning was the industry standard. A *New York Times* writer reporting on a major trade show that year observed that the single-control feature no longer attracted special attention because it was standard on most models.[22]

Advertisers still had to overcome many potential customers' unfamiliarity with even the simplest radio. Improvements such as those described above allowed them to develop the idea that using radio was, as one ad proclaimed, as "simple as calling a telephone number." Even women could manage it. The same maker boasted of another model that its "selectivity appeals to everyone—and to the women folks in particular. You can . . .

tune in any local station you wish and not be bothered with interference from the others." Unlike their male counterparts, who had learned to engage skillfully with the technology, women were expected to be more passive users. Increasingly during the twenties factory-built models, especially the more expensive ones, were linked to feminine settings. Employing deliberately different sales strategies for men and women helped move radio beyond its narrow hobby niche.[23]

The creation of a new social context for radio was as important to increasing its popularity as any innovation in the laboratories of RCA, GE, or Atwater Kent. In contrast to the wireless, the modern preassembled radio reinforced middle-class domestic conventions by encouraging women to stay safely inside the confines of home and traditional roles. Receivers made in 1922 and 1928 might perform the same technical duties of transforming electromagnetic signals into intelligible sound, but the social contexts in which they commonly were portrayed differed greatly.

Manufacturers recognized the importance of encouraging women users, most of whom had little experience with electrical equipment. Of course they also sought male consumers who had avoided hands-on experience during the boom years of wireless.[24] In general, however, the advertisements they sponsored and the advice their agents dispensed to retailers made clear that women were the crucial audience. With these strategies they effectively reinvented radio as a device not merely accessible to but particularly suited to female listeners.

Despite their cultivation of female radio users, advertisers did not lose sight of men altogether. In fact, much of their copy was created largely for male eyes. If radio was a household device designed specifically to be accessible to women, why was advertising was aimed almost exclusively at men? Advertisers believed that women, as traditional guardians of domesticity, would be interested in radio as furniture and a source of family entertainment. They also believed, with good reason, that it was predominantly men who bought receivers. Both constituencies could be addressed. Radios were made attractive to women so that they would encourage men to buy them. The woman of the house could instigate the purchase, although a man would actually carry out the transaction. Or if radios appeared to be what men thought women wanted, then men could experience consumer desire themselves and initiate the purchase directly.

If radio was to please the middle-class housewife and assist her in bringing culture to the home, it would first have to be sold to her husband. In the absence of sales figures distinguishing between male and female customers, we must draw inferences from other sources, but there is both negative and positive evidence on behalf of this roundabout strategy. First is the relative absence of radio advertising from major publications catering to female audiences, such as *McCall's, Ladies' Home Journal,* and *Better Homes and Gardens.* Although the makers of the Victrola, a well-established music technology, advertised regularly in these outlets, Atwater Kent was the only receiver maker to venture the occasional spread in these feminine pages, and it did not do so until 1926.

This reluctance cannot be attributed to uncertainty whether readers could afford the latest consoles.[25] When it did use these outlets, Atwater Kent showed off the middle and top of its line, ignoring lesser models. In *Good Housekeeping* copy, for example, contemporary writer Wallace Irving attested to the virtues of an Atwater Kent five-tube compact receiver, a mid-price set at $80. Immediately thereafter the makers of Pooley cabinets joined forces with Kent, urging readers to "acquire a lasting pleasure" for $295. Had readers' spending levels been a concern, any lower-priced competitor could have stepped into the advertising void left by Kent and Pooley in the pages of women's magazines. Newspapers were the most common medium for advertising radios, and copy there remained directed at male readers for most of the twenties. In the *New York Times,* for example, radio ads ran in Sunday sections given over to automobiles and airplanes, two other predominantly male concerns. Not until the early thirties did the radio section and accompanying ads find a regular home in the cultural pages of the *Times.*[26]

Trade publications made clear that while the ideal radio user might be a woman listening while doing housework or calisthenics, the typical customer remained male. Ad copy should attract men's eyes, dealers were advised. "The female of the species is more attractive than the male," the editors of *Radio Merchandizing* told their enterprising male readers. Taking a hint from publishers who lured readers by placing "the wily charms of the gentler sex" on magazine covers, the wise retailer would "make the dominant note of his window display distinctly feminine." The editors urged subscribers to emulate the publishing industry's success by placing female mannequins "reclining in languorous positions" in storefront displays. His eyes drawn by these models, the prospective customer would pause to appraise the goods. He would see the putative benefits to be gained not by him but by his wife. This too was according to plan. If the dealer prepared his display window to resemble "milady's boudoir," he could indicate on a printed card that "the lady is getting her daily beauty hints or that she is being entertained" in the comfort of her bedroom. The promised payoff for this elaborate setup would be "more men outside that window than ladies" and subsequent higher sales inside.[27]

Differentiation between male purchasers and female users appears frequently in advice on retailing strategies. Suggested sales campaigns, editorial advice, and contrived vignettes made this abundantly clear: men were expected to come into the store, but the enterprising retailer would stock merchandise appealing to women's sensibilities and plan sales campaigns accordingly.[28]

Advice to dealers is important evidence because radio retailing remained a predominantly local enterprise through the twenties. While other household items, such as electric irons, enjoyed brisk sales through national or regional chains, radios remained concentrated at individual electric supply shops or small local companies, which were more likely to rely on local newspaper advertising and to write their own copy than large conglomerates. This is one reason why it was so difficult to domesticate radio fully. As late as 1930 this important household item had not been integrated into the chain store, an institution important to our understanding of twentieth-century consumerism. Only one-fifth of sales in that year were rung up by national chains, while single-store independents accounted for more than 60 percent of all sales. In contrast, fully half of other household appliances were sold through national concerns.[29]

Radio was not a high-priority item for department stores, palaces of consumption catering to women. Deals for receivers were struck man-to-man across the counters of electrical supply, radio, music, and hardware stores. *Radio Merchandizing* cajoled its readers to understand the "Woman Market" and to fathom "the Buying Habits of Women" as a way to increase sales, especially to married men. The magazine advised that appealing to a woman's presumed interests would likely get her "boosting for a radio outfit."[30]

At the same time the industry clung to the view that because of its technical nature, radio was best handled by experienced electrical shops or specialty stores, where women generally did not venture.[31] This was a self-serving position for dealers to take yet not one that can be dismissed. However cleverly disguised in a fine wood cabinet, radio remained for retailers and prospective buyers a complicated piece of equipment whose virtues could be demonstrated best by someone both expert and male. Its male-dominated ancestry was not easily discarded. Radio might please the beauty queen shown in one ad perched atop a receiver, who pledged that "you can't be beautiful unless you're happy," but she would have to depend on a man for that kind of happiness. A 1926 ad run in *Life* brought this point home. Addressed "To the Girls—Workers All," the copy for Grebe Synchrophase receivers asserted that all women exchanged labor for material gain. "Even when little Gloria Staholm slips an arm around Daddy's neck and playfully pulls his ear, *she's* working—Dad. And why not? How else to gain those things that make life pleasant?" The ad then confidentially suggested that "one of the easiest jobs, girls, is to persuade Dad to buy a Synchrophase." The trick was to motivate her to ask him for it, or better still, train him to anticipate her desire.[32]

Cultivating female users as a route to the wallets of male purchasers required putting radio in a feminine physical context. Selling radio as merely an isolated electro-mechanical device would not suffice. Gone were the pictures of Boy Scouts using wireless at camp. Now advertisers portrayed the receiver fitting into and improving the home. By depicting it as part of a calm and elegant domestic scene they effectively created a spatial context where none had existed. Because radio implied potentially disruptive changes to family routines, promoters took pains to harmonize it with the household setting. They often depicted it alongside lamps, portraits, and other familiar domestic objects. Although radio remained the subject of these illustrations, it fit harmoniously with the domestic environment constructed around it. The radio room often appeared without occupants, emphasizing how a receiver's design and craftsmanship ingratiated it into familiar domestic space.

When occupants did appear, typically both men and women were shown, often sharing the radio space with a small child, almost invariably a girl. They listened with interest and pleasure but not with such intensity that they could not simultaneously converse, read, or entertain. These scenes reinforced the idea that radio was no longer a gadget for men and boys and also suggested how middle-class consumers could change the ways they used living space. That space, and presumably the patterns of family activity flowing through it, could accommodate a new presence. Through innumerable print advertisements most of the major receiver manufacturers (including Ultradyne, Atwater Kent, Crosley, Kennedy, and Magnavox) presented their vision of radio in an idealized setting. With only a slight adjustment of traditional furnishings and family habits, these ads implied, the radio could be integrated into the background of middle-class living space and life. Since engineers had already redesigned receivers to make them less demanding of technical skill, it was the advertiser's job to suggest how the home's appearance and function could be improved by the "tastefully unobtrusive" incorporation of an Ultradyne model into the home space occupied by a well-dressed couple. "Radio needn't disturb any room," an Atwater Kent ad promised, but it could create the sort of home a man might want, or, more accurately, the kind a man might desire his wife to want.[33]

In order for radio to be successfully domesticated, women had to want it in the home. But since advertisers assumed that a man would make the actual purchase, he had to want it too. Suggesting how radio might help sustain marital harmony and insulate his family from immoral influences was one way to encourage that. This leads to the final and perhaps most important point to be made about radio's gendered domestication: the conflicted relationship that radio's promoters saw between it and modernity. In the enthusiastic rhetoric of radio boosterism, by no means all of which was created by advertisers, the new medium seemed to define the modern age. It was electric, its effects upon early listeners electrifying. It seemed to represent science bending the laws of nature to humankind's intellectual and cultural welfare. It audibly demonstrated the insights of great inventors brought to fruition under American capitalism. It cut through time and space, bringing the stimulation of theaters, concert halls, and sporting arenas to everyone. The enthusiastic ad copy would lead one to think that this was the future Americans had been dreaming about.

Advertisers kept one foot in the future and the other planted firmly in the past. They understood that the new medium could appeal to men seeking refuge from the rougher aspects of modern life. Many city dwellers feared the influences of jazz and swing, dance halls, cabarets, and Hollywood movies. Old rules governing social contact between the sexes had been broken by modern entertainments; new activities enticed young women into close, sometimes immediate contact with men. The auto's ability to remove couples from watchful eyes, dark movie theaters, crowded dance halls, amusement park devices that threw riders into each other's arms—all these challenged the notion of separate spaces for men and women and mocked middle-class modesty.[34] In contrast to these amusements, radio could entertain the whole family within the safety of the home. By providing a way for women to experience the world's pleasures without actually stepping out, it seemed to shore up genteel traditions against the modernist hazard.

Advertisers recognized the concern among white middle-class males that American culture had become debased, endangering family and moral traditions. Trade journals addressed these fears proudly and directly. One urged dealers preparing for the Christmas holidays to "get out a good circular letter to the Daddy customers on your prospect list" emphasizing the "'keep the youngster home' benefits" of radio as opposed to the benefits of more traditional gifts. The editors of a trade association journal bluntly declared that radio meant "that young people need not leave the home in search of some place where 'something is going on.'" As a result, "family life is no longer in danger because of the influence of the automobile and the motion picture."[35] Other household technologies from this era reinforced traditional gender relationships. But unlike the washing machine or the electric iron, which appeared, however deceptively, to liberate women from generations-old tasks, radio was specifically promoted to a male market based on its alleged ability to sustain the past by persuading women to remain at home.[36]

Urged to "tell the story of radio to the whole family," retailers adopted ads emphasizing traditional female roles in the home. While growing numbers of women sought outside employment, one suggested strategy was to show a young woman happily listening to a radio while sewing, over the caption "While You Work." An ad like this, suggesting the "home touch of working while listening in," would, the magazine promised, "instantly appeal to the lady of the house" and presumably to the man likely to purchase the set.[37] Other advertising offered subtle variations on this theme. Magnavox, for example, promised to improve married life by easing the tension between the tradition of staying in and the titillation of going out. A 1923 ad shows a couple relaxing intimately before their receiver while a nanny leads the children off to bed. The moral of this tableau is that husband and wife are "Pals again. Never a dull evening in the home." A 1926 ad makes the point with more subtlety, promising a man working out household expenses that a radio "Cuts Your Entertainment Cost in Half." It suggests that he could control both the activities of his family and the cost of its consumption habits by purchasing a consumer technology.[38]

Numerous other illustrations show idealized white affluent domesticity in which children are happily united with mothers, and husbands with wives, before a receiver. Befitting their passive role, the female figures are invariably seated, whereas the males may be standing. The emphasis on this definition of femininity was no accident, of course, given the fear that the flapper's penchant for novelty and traditionally masculine activities, from driving to smoking to voting, might dissolve family bonds. In fact one of the few exceptions to the familial motif shows several fashionable women, thoroughly modern in appearance, enjoying tea by their FADA receiver. For those who might miss the point, the copy explains that when installed at home, the FADA's "chaste" cabinet "harmonizes with any interior." That is the text. But the implied message is that these modern and presumably chaste women were happily harmonized with domestic life by their receiver. By its very presence the receiver seemed to free the home from the threat of sexual liberation.[39]

This story would not be complete without a note of caution. Radio clearly occupied a different place in the home in 1932 than it had in 1922. Its capabilities, control mechanisms, and outward appearance had all been redesigned to accommodate the change from hobby device to instrument of culture and social control. We do not know exactly how individual families used the device or to what degree they put advertisers' suggestions into practice, yet radio's transformation was neither as complete nor as seamless as manufacturers wished. However much ad copy might boast that it had been transformed from yesterday's "scientific marvel" into the means of bringing "the most thrilling interest and enjoyment within reach of the average American home" today, it was not divested of its past quite so rapidly. Like every technology, it still carried evidence of its origins.[40]

David Sarnoff had been right. By the mid-thirties more American families had radios than either telephone service or running water. The industry's sales strategies had effectively capitalized on the excitement of radio as well as on public enthusiasm for the future and unease about the present. Although radio had not become a truly universal form of communication, it occupied a place in the homes and daily lives of a large majority of families. The second part of Sarnoff's prediction obviously has not come true. It is questionable whether at any point in its history radio could be said to have functioned as "a cleanser for the mind." However, many people both inside and outside the broadcasting industry thought that it should, and Sarnoff's vision held considerable influence in debates about radio program content, particularly at the National Broadcasting Company (NBC), the network owned by Sarnoff's RCA. When manufacturers of early television sets adopted sales strategies very similar to those used thirty years earlier for radio, it was inevitable that the predictions and debates about the effects of broadcasting in the home would be repeated. In the late twentieth century, with millions of radio and television sets enabling us to consume broadcast programs as routinely as water and soap, we have not decided what these technologies should do, and for whom. The novelty of such issues having long worn away, we merely address them less directly. Old questions about the place of entertainment technologies in American family life take on new forms in political wrangling over broadcast regulation, signal scrambling, the so-called V-chip, and access to the Internet. The boundaries between public and private behavior and between purported male and female attributes have been considerably redrawn since 1922, but we Americans are still anxiously seeking the technology that will make tensions between the sexes, between generations, and between subcultures simply go away.

Notes

I wish to thank Barbara Becker, Cathy Brill, Ian Corbyn, Stuart Leslie, Arwen Mohun, and participants in the Hagley Library's "His and Hers" symposium for comments on drafts of this chapter.

1. Sarnoff address to fifth annual convention of the New York State Association of Electrical Contractors and Dealers, 13 June 1922, Biographical Reference File 132 (Sarnoff), folder 1922, Broadcast Pioneers Library at the University of Maryland, College Park. Sarnoff was promoted from commercial manager to executive vice president of RCA in 1922. The standard biography of Sarnoff is Kenneth Bilby, *The General: David Sarnoff and the Rise of the Communications Industry* (New York: Harper & Row, 1986). A critical treatment of Sarnoff's early career is Thomas S. W. Lewis's triple biography, *Empire of the Air: The Men Who Made Radio* (New York: Edward Burlingame, 1991).

2. When the U.S. Bureau of the Census first enumerated households with running water, in the 1940 census, it reported 19,174,344 residences with a private bathtub, toilet, and sink. This is slightly less than the number of families estimated to own radio sets seven years earlier (U.S. Bureau of the Census, *Census of Housing*, vol. 1 [Washington DC: GPO, 1943], 8; *The Statistical History of the United States from Colonial Times to the Present* [Stamford CT: Fairfield, 1965], 491). In 1940 the National Association of Broadcasters (NAB) estimated that there were 44 million receiver sets in 28 million homes. The NAB numbers, like other industry estimates, cannot be independently verified but appear consistent with manufacturing data. By 1929 commercial production of receivers (4.98 million) outstripped by many times that of bathtubs (0.94 million) (U.S. Bureau of the Census, *1930 Census of Distribution* [Washington DC: GPO, 1933], 90, 110; the NAB estimate is cited in Paul F. Peter, "The American Listener in 1940," *Annals of the American Academy of Political and Social Science* 213 [1941]: 2).

3. Margery Griffin, "The Radio Widow's Plaint," *Wireless Age* 12 (Dec. 1924): 25.

4. Susan J. Douglas, *Inventing American Broadcasting, 1899–1922* (Baltimore: Johns Hopkins Univ. Press, 1987), ch. 6 and 301–3; Susan Smulyan, "The Rise of the Radio Network System: Technological and Cultural Influences on the Structure of American Broadcasting," *Prospects, An Annual of American Cultural Studies* 11 (1987): 105–17.

5. Jack Binns, foreword to Allen Chapman, *The Radio Boys' First Wireless* (New York: Grosset & Dunlap, 1922), v–vi. I am grateful to Deborah Hirschfield for bringing the *Radio Boys* series to my attention. On Binns's role in the *Republic* accident see Douglas, *Inventing American Broadcasting*, 200–202.

6. Douglas, *Inventing American Broadcasting*, 298–307. The titles appearing in 1922 were *Popular Radio* (New York, to 1928), *Radio Age* (Chicago, to 1928), *Radio Broadcast* (Garden City NY, to 1930), *The Radio Dealer* (New York, to 1928), *Radio Programs Digest Illustrated* (Chicago), *Radio Merchandising* (New York), and *Radio World* (New York). The others from this period were *Radio Amateur Callbook Magazine* (Chicago), *Radio Call Book and Technical Review* (Chicago), and *Radio Listener's Guide and Call Book* (Chicago). Titles and publication information were compiled from serials catalogues at the Library of Congress.

7. Although "The Radio Woman" sometimes offered useful hints or stories about women who had solved technical problems, it more often concerned matters of using radio for self-improvement or creating a more pleasant home (see "Women Inspect Bulbs at U.S. Laboratory," *Radio World* 5 [3 May 1924]: 15; "A Transcontinental Reflex," ibid., 5; and "Woman Helps in Home Radio Problems," ibid., 19 Apr. 1924, 15). In her dissertation on the role of the daily press and popular magazines in creating audience desire for commercial broadcasting Laura Pelner McCarthy cites several instances in which women or girls were presented as knowledgeable radio users early in the decade. The scope of this dissertation does not encompass either receiving hardware or paid advertising (McCarthy, "The Limit of Human Felicity: Radio's Transition from Hobby to Household Utility in 1920 America" [Ph.D. diss., University of Florida, 1993], 90–93).

8. Douglas, *Inventing American Broadcasting*, xxviii.

9. Smulyan, "Rise of the Radio Network System," 105–12.

10. To my knowledge, the only other study of receiver advertising that raises these questions is Thomas W. Volek, "Examining Radio Receiver Technology through Magazine Advertising in the 1920s and 1930s" (Ph.D. diss., University of Minnesota, 1991). Volek's description of the change in the way advertising portrayed radio, first as a hobby device and later as domestic furniture, complements my own. Volek's dissertation presents a wealth of sources. As its title suggests, these are advertisements drawn from magazines; advertisements from newspapers and trade publications are not included. Volek acknowledges the differential gender and class appeals of ads but ultimately subsumes these within the consumer demands of an unspecified "public" without suggesting underlying cultural reasons for their success (see esp. chs. 5 and 6). Another informative but uncritical source is Leslie J. Page Jr., "The Nature of the Broadcast Receiver and Its Market in the United States from 1922 to 1927," *Journal of Broadcasting* 4 (1960): 174–82. An encyclopedic collection of technical specifications and advertisement reproductions devoid of historical interpretation is Alan Douglas, *Radio Manufacturers of the 1920s*, 3 vols. (Vestal NY: Vestal Press, 1988–92).

11. Westinghouse put station KDKA on the air in Pittsburgh in 1920 and recognized the amateurs listening to the station as "simply the forerunners of a much

larger market for radio receivers." AT&T established New York station WBAY (later WEAF) in 1922 as the anchor of what it hoped would be a chain of advertising-supported ("toll broadcasting") transmitters linked by telephone company wires (Douglas, *Inventing American Broadcasting,* 300; William Peck Banning, *Commercial Broadcasting Pioneer: The WEAF Experiment, 1922–1926* [Cambridge: Harvard Univ. Press, 1946], ch. 5).

12. Eric Barnouw, *A History of Broadcasting in the United States: Vol. 1, A Tower in Babel* (New York: Oxford Univ. Press, 1966); Susan Smulyan, "Rise of the Radio Network System." On audience behavior see Lawrence W. Levine, "The Folklore of Industrial Society: Popular Culture and Its Audiences," *American Historical Review* 97 (December 1992): 1369–99; Susan J. Douglas, "Notes Toward a History of Media Audiences," *Radical History Review* 54 (1992): 127–38; Lizabeth Cohen, "Encountering Mass Culture at the Grassroots: The Experiences of Chicago Workers in the 1920s," *American Quarterly* 41 (1989): 6–33.

13. On press coverage of the annual exhibition see, e.g., Orrin E. Dunlap Jr.'s description of the third Radio World's Fair, "Exhibits Disclose Radio's Advance," *New York Times,* 12 Sept. 1926, sec. 12, p. 1. The seasonal nature of radio retailing has been unremarked except for a brief mention in Page, "Nature of the Broadcast Receiver."

14. In an analysis of the 1930 census Herman S. Hettinger and Walter J. Neff found that although the wealthiest families accounted for but 2 percent of all radio-owning households in 1936, their probability of owning more than one set increased their total share of the market. The authors estimated that 15.5 percent of families owning radios owned more than one set and that virtually all of these families belonged to the middle- and upper-income brackets (Hettinger and Neff, *Practical Radio Advertising* [New York: Prentice-Hall, 1938], table 3).

15. Oversize Book 185 [Eveready], N. W. Ayer Advertising Agency Records, Archives Center, National Museum of American History, Smithsonian Institution, Washington DC (hereafter N. W. Ayer, NMAH).

16. *Radio Broadcast* 7 (Nov. 1925): 67.

17. Ibid.

18. Orrin E. Dunlap Jr., "Progress of Radio Revealed at Show," *New York Times,* 18 Sept. 1928, 32.

19. Book 17 [FADA Radio], 1924 folder, N. W. Ayer, NMAH.

20. Keith Henney, *Principles of Radio,* 5th ed. (New York: John Wiley & Sons, 1945), in which ch. 16 provides a basic, nearly contemporary explanation of receiver electronics; "Operating the Dials Is an Exact Art," *Radio World* 5 (19 Apr. 1924): 10.

21. In addition, some receivers had a separate control for adjusting the antenna coil (Arthur P. Harrison Jr., "Single-Control Tuning: An Analysis of an Innovation," *Technology and Culture* 20 [1979]: 296–302).

22. Ibid., 308–10, 315–21. The most popular new feature of the late twenties and early thirties was the ability to use household alternating current instead of batteries (Dunlap, "Progress of Radio").

23. Both examples appear in Book 17 [FADA Radio], 1924 folder, N. W. Ayer, NMAH.

24. One early guide emphasized that the successful radio store must distinguish between the male radio "nut" and the novice "fan." The former already understood radio principles and knew what he wanted. Salesmen were likely to get tied up in lengthy technical conversations with him at the expense of tending to the "fan," who was more open to expensive suggestions (*How to Retail Radio* [New York: McGraw-Hill, 1922], ch. 5).

25. Magazines whose audiences included both men and women, such as the *Saturday Evening Post,* ran considerably more radio advertising than did strictly women's publications. The difference between radio and other consumer items is underscored by the fact that women's magazines carried numerous advertisements for automobiles, another modern technology presumably of particular interest to men.

26. See Kent's ad in *Good Housekeeping* 26 (Feb. 1926): 82; and Pooley's in ibid., Mar. 1926, 82. There appear to have been no definitive rules governing the placement of radio material in the *Times.* In 1927 radio was mentioned regularly in the drama, music, and art sections; the following year, however, it returned again to the auto, aviation, and editorial pages.

27. "The Female of the Species Is More Attractive Than the Male," *Radio Merchandizing* 2 (Mar. 1923): 30.

28. One such example is "Mrs. Old Timer Takes Up Radio," ibid. 3 (July 1924): 20.

29. U.S. Bureau of the Census, *1930 Census of Distribution,* vol. 1 [Retail Distribution], (Washington DC: GPO, 1933), 30 and table 6. On department stores see Susan Porter Benson, *Counter Cultures: Saleswomen, Managers, and Customers in American Department Stores, 1890–1940* (Urbana: Univ. of Illinois Press, 1988). On the relationship between mass consumption and mass retailing see Susan Strasser, *Satisfaction Guaranteed: The Making of the American Mass Market* (New York: Pantheon, 1989). Cohen, "Encountering Mass Culture at the Grassroots," is an important examination of consumer culture at the level of neighborhood retailers.

30. *Radio Merchandizing* 4 (June and July 1925); quotation from ibid. 3 (Apr. 1924): 63.

31. At least one publication recognized an alternative to this situation. *How to Retail Radio,* a guide printed by the publisher of *Electrical Merchandising* magazine, cautioned that the "market will not be limited to the men; women and girls are buying." These prospective customers needed special care to make them comfortable in

the radio store. The book suggested that dealers create a listening area in a quiet corner and furnish it to simulate the familiar home environment. Another option was to offer demonstrations in the customer's own home (*How To Retail Radio,* ch. 5).

32. *Radio World* 5 (14 Apr. 1924): front cover. On rare occasions how-to magazines such as *Radio World* printed articles about women building their own receiver sets from components. The text made clear, however, that these women were to be regarded as novelties (Grebe ad in Volek, "Examining Radio Receiver Technology," fig. 68; it originally appeared in *Life,* 21 Jan. 1926, 25). Volek interprets this example as an appeal to working-class consumers but overlooks the gender implications of the text.

33. Ultradyne ad in *Radio Broadcast* 7 (Nov. 1925): 67; Atwater Kent ad in Volek, "Examining Radio Receiver Technology," fig. 65, originally in *Ladies' Home Journal,* Dec. 1925, 179.

34. The literature on early- to mid-twentieth-century commercial entertainments is stimulating, large, and growing. Standard studies of movies, amusement parks, dance halls, and nightclubs are, respectively, Lary May, *Screening Out the Past: The Birth of Mass Culture and the Motion Picture Industry* (New York: Oxford Univ. Press, 1980); Kathy Peiss, *Cheap Amusements: Working Women and Leisure in Turn-of-the-Century New York* (Philadelphia: Temple Univ. Press, 1986); John Kasson, *Amusing the Million: Coney Island at the Turn of the Century* (New York: Hill & Wang, 1978); and Lewis A. Erenberg, *Steppin' Out: New York Nightlife and the Transformation of American Culture, 1890–1930* (Westport CT: Greenwood Press, 1981).

35. *Radio Merchandising* 4 (Nov. 1924): 36; *RMA News,* 25 Oct. 1928, 11, published by Radio Manufacturers Association, ser. 217, box 206, George H. Clark Radioana Collection, Archives Center, NMAH.

36. Ruth Schwartz Cowan, *More Work for Mother: The Ironies of Household Technology from the Open Hearth to the Microwave* (New York: Basic Books, 1983). A quantitative assessment of the effects of modern appliances on the time spent doing housework is Charles A. Thrall, "The Conservative Uses of Modern Household Technology," *Technology and Culture* 23 (1982): 175–94.

37. *Radio Merchandizing* 3 (Apr. 1924): 63.

38. Stewart-Warner ad in Volek, "Examining Radio Receiver Technology," fig. 84, originally in *Saturday Evening Post,* 13 Nov. 1926, 54.

39. Book 17 [FADA Radio], 1924 folder, N. W. Ayer, NMAH.

40. Magnavox in *Radio Broadcast* 1 (May 1922): 75.

Six

Cinderella Stories
The Glass of Fashion and the Gendered Marketplace

Regina Lee Blaszczyk

Few Western women experience girlhood without hearing, reading, or seeing the ancient fable of Cinderella, the soot-covered, suffering servant who finds happiness as a princess with the aid of magic and, more important, a portfolio of fashionable consumer goods. Forbidden to attend a royal ball by her wicked stepmother, the impoverished maiden labors in the kitchen while her ugly stepsisters bedeck themselves for the fete. Following their departure, the miserable Cinderella is visited by her generous fairy godmother, who outfits the beauty with costume, coach, and livery designed to capture royal attentions. After several blissful evenings dancing with the enthralled prince, Cinderella dashes from the ballroom to her carriage to meet the gift-giving godmother's midnight curfew, losing a crystal shoe in her final flight. Ultimately this discarded dancing slipper is the key to the lovers' reunion, for Cinderella's glass slippers are crafted from a precious material and precisely calibrated to hold only her feet. The infatuated prince scours his kingdom for his true love, finally locating his bride-to-be, Cinderella, the sole girl whose foot fits into the glittery glass slipper. United, the two live happily ever after, or so the story goes.[1]

For centuries the Cinderella story has offered girls and women a paradigm for realizing emotional fulfillment and achieving financial security through matrimony and attendant domesticity. Yet this tale is more than a prescription for culturally sanctioned gender roles; it also reveals much about the centrality of material goods—including glass—in building individual identity and negotiating interpersonal relationships. Without question, Cinderella's transformation from a kitchen wench to a dazzling belle depended on a fairy's gifts, and salvation from her evil keepers required the goodwill of a determined male protagonist, the prince, who epitomized a future of marital bliss. But Cinderella captured her prince's attentions and secured his devotion by aptly using her godmother's bequest, an ensemble of fanciful consumer products. Cinderella's luxury vehicle and her exquisite evening clothes, including her one-of-a-kind glass slippers, were fashioned to make a queenly impression. Only with the aid of this constellation of things was Cinderella's natural beauty amplified into the courtly good looks that caught the prince's eye, turned his head, and held his gaze. The material message in this rags-to-riches romance is clear; objects are levers that women can use to set agendas and to achieve ends within the gendered realms of courtship and marriage.

Like all great fables, the Cinderella tale operates on many metaphorical levels, but its object lessons are especially pertinent to scholars seeking to understand the underpinnings of cultural production in twentieth-century America. As in the Cinderella romance, gender politics figured boldly in the culture and economy of modern America, infusing interactions between manufacturers, retailers, and consumers of household furnishings and accessories with a creative dynamism. As consumers, women were primarily responsible for purchasing and caring for their families' domestic furnishings and accessories, including household glassware. In many respects they were Cinderellas seeking the good life by using certain types of objects to mediate gender relationships as prescribed by mainstream culture and tempered by ethnic, racial, and economic subcultures. Through the activities of selecting, purchasing, and using those artifacts, consumers forcefully challenged producers' capabilities and expectations. In turn, managers in manufacturing, retailing, and advertising continually reassessed the marketplace, designing and redesigning commonplace domestic accessories to meet women's fancies. Just how glass manufacturers imagined the tastes and met the expectations of modern Cinderellas, giving form to the market basket of household glassware, is the subject of this chapter.[2]

Successful glassworks managers led their firms through the quagmire of the gendered marketplace by straddling the spheres of production and consumption. To accomplish this acrobatic feat, they depended on the feedback of those people who best understood women's tastes, that is, on professional *fashion intermediaries*, who earned their livelihood by deciphering consumers' desires. In some respects these fashion liaisons were elves who helped manufacturers fulfill their roles as surrogate fairy godparents in real-life Cinderella stories. At the Fostoria Glass Company in Moundsville, West Virginia, managers relied on the input of consumer intermediaries such as product designers, advertising executives, retailers, and home economists as they labored to retain their firm's foothold in a shrinking market for formal tableware and stemware. During the interwar and postwar years culturally sanctioned gender roles influenced consumers' tastes and perceptions and decisively shaped the market for products such as Fostoria glassware.[3]

In many respects Fostoria typified American glass factories in the upper Ohio River valley during the early- to mid-twentieth century. Established in 1887 near Toledo, the firm moved to Moundsville, West Virginia, when natural gas supplies were exhausted at the century's end. In 1901 William A. B. Dalzell became the company's president; the Dalzell family and managers drawn from the local population controlled the glassworks until its closing during the 1980s. Like many midwestern glassmakers, Fostoria was a master of a manufacturing style that some scholars and business theorists have termed flexible production. At Fostoria and other midwestern factories highly skilled craftsmen made glassware in part by hand, using small, portable presses kept near furnaces. As a result, these factories were known in the industry as "hand-glass plants." This type of labor-intensive production allowed Fostoria's managers to make considerable quantities of glass in many shapes and patterns to satisfy consumers' highly variable tastes. With this approach to manufacturing, the Dalzells had built Fostoria into one of the nation's prominent makers of glass tableware and stemware by World War I.[4]

Fostoria's ascendancy to leadership in the hand-glass industry was based on Dalzell's skills in imagining consumers' product expectations as configured by the dominant American culture. During the nineteenth and early twentieth centuries mainstream culture sanctioned middle-class values and behavior, with the home and its furnishings functioning as consumers' major vehicles for expressing class affiliation and individual style. As guardians of the private sphere, women were charged with selecting, arranging, and orchestrating the use of household objects. Among the hallmarks of middle-class status was the ritual of genteel dining, which required the use of a spectrum of accouterments, from water goblets to salt cellars, and a small staff of servants who helped to prepare and serve the evening meal. A hostess's discrimination in selecting and arranging various utensils—the degree to which her china, glass, and silver harmonized with surroundings to create a pleasing environment—conveyed much about a family's social position and a woman's personality. The right artifacts reflected flatteringly on their owners, just as Cinderella's attire, including her crystal slippers, leaves a positive impression on the prince.[5]

Thanks to women's highly varied tastes, the market for dining accessories divided into untold segments, and manufacturers such as Fostoria responded by developing products in a range of styles and prices. But one thing about this business remained constant; manufacturers designed pressed glassware to imitate the look of costly crystal. In Cinderella's day the state of glassmaking technology circumscribed production in terms of quantity and quality, so that clear, colorless, transparent, and brilliant glassware was an expensive luxury found exclusively on the tables of the rich and famous. From the Renaissance to the Enlightenment, European aristocrats and nobles exchanged gifts of extraordinary glassware, often made by Venetian craftsmen and richly ornamented with engraved or cut motifs, and used these showy goods in elaborate, festive meals. This elite legacy of presentation, display, and ceremony cast a long shadow, for even after the American innovation of pressing democratized glassware consumption during the early nineteenth century women continued to exalt crystal as the epitome of the glass medium. While only the wealthiest Americans could afford costly handmade crystal, middle-class consumers stocked their cupboards with pressed glassware made in copycat cut-glass forms for serving or drinking. In the dim light of a formal dinner bits of highly refractive imitation cut glass sparkled, diamondlike, lending a celebratory aura to a genteel evening meal.

After World War I, Fostoria and other glassmakers grappled with major demographic and social shifts, including the so-called servant problem and national prohibition, that were recasting the demand for consumer products. During the 1920s several factors, from the expansion of factory and office jobs for working-class women to nativist intolerance of immigrants from southern and eastern Europe, contributed to middle-class perceptions of a servant shortage. The adoption of the Eighteenth Amendment, which outlawed alcohol production in 1920, reshaped cooking, dining, and drinking practices, circumscribing the preparation of French dishes that either contained or were eaten with wine, sherry, or liquors. Without good help and the appropriate ingredients American homemakers found it difficult to practice the ritual of formal dining. Concurrently, agents of the emerging consumer culture—moving pictures, radio programs, and mass magazines—engendered consumers with an appreciation for glamorous, urban living and colorful, modern fashions. In this context American women began to reevaluate consumption habits inherited from their grandmothers, reconfiguring their visions of what constituted the ideal meal and the perfect ensemble of dining accessories.[6]

To be sure, consumers still acquired china and glassware, for these goods had not lost their symbolic functions. Tableware could signify emotions and anchor memories: a dinner service made from fine white porcelain given as a wedding present might epitomize a bride's virginal purity and guileless character, or a bonbon dish fabricated in pressed glass might remind a newlywed of a relative living in a faraway place. Many consumers continued to identify table accessories with food, so that otherwise ordinary objects—dishes, bowls, and tumblers—were emblematic of sustenance in its many forms. As such, sets of matching porcelain, glassware, and silverware, artistically exhibited in fancy china cabinets, connotated many women's faith in the sanctity of the traditional life-sustaining institution, marriage. But even though symbolism weighed heavily in many consumers' minds, the declining usefulness of enormous tableware began to erode the glassware market as it had existed for generations. Women still liked to set fancy tables, but they planned menus and selected equipment that met their new-found needs as modern homemakers. The china service remained the backbone of a hostess's dining outfit, but the mock-crystal glass set, replete with olive dishes, celery holders, wine goblets, and other specialized forms, was relegated to peripheral status, a relic of the multicourse meal of bygone days.[7]

In this context Fostoria's managers created a product line geared toward accommodating women's unrelenting passion for glassware, their invariable yearning to keep up appearances, and their compelling desire to modulate Victorian dining customs to suit modern times. Beginning in 1925 Fostoria unveiled a unique line of glass tableware: dinner services, breakfast sets, luncheon sets, and tea services—all in pale shades of green, blue, pink, purple, and yellow. With these modestly priced, stylish products the firm expected to wrest a share of the profitable trade in fine china services away from porcelain manufacturers such as Lenox, Spode, and Wedgwood. To facilitate sales, department stores and china shops carried Fostoria glassware in open stock, a type of installment plan that allowed consumers to purchase items when they wished. Like auto manufacturers seeking to gain competitive advantage in the style-conscious 1920s, Fostoria reoriented its product lines to meet consumers' changing lifestyles and expectations rather than to manipulate their tastes.[8]

With the introduction of exotic tints such as orchid, dawn, topaz, and azure Fostoria's managers joined the ranks of American manufacturing executives who acknowledged that color, with its powerful effect on mood, might be used to sell consumer goods. The popularity of color psychology, the publicity surrounding the Exposition Internationale des Arts Décoratifs et Industriels Modernes in Paris in 1925, the development of new paints, lacquers, and other compounds by the nation's expanding chemical industry, and the initiation of a Color in the Kitchen movement by New York retailers had spurred a demand for chromatic housewares. Companies making kitchen ranges, linoleum floor coverings, and plumbing fixtures made colorful goods and advertised them in popular magazines. At Fostoria, Dalzell engaged the expertise of his factory chemist to create batch formulas for pale, transparent glasses. With recipes in hand, Dalzell called on the expertise of professional designers, charging these fashion intermediaries with the task of determining consumers' taste preferences in the context of the chromatic revolution.[9]

Crystal's legacy cast a long shadow. Fostoria's in-house designers, Edgar M. Bottome and George Sakier, faced the challenge of engaging the gauntlet of style to fracture this heritage without alienating potential customers or obviously seeking to control their tastes. Older consumers were critical to Fostoria's sales, for mothers, grandmothers, and great-grandmothers versed in the Victorian ritual of fine dining often purchased the accouterments of gracious housekeeping for their engaged offspring. Younger consumers cared less about formality, but they brought an intensified appreciation of fashion, including the color vogue, to their shopping choices. The trick to appealing to this multigenerational audience rested in employing orthodox aesthetics, in replacing the repertoire of cut-glass motifs with equally conservative conventions that appealed to old and young alike. To do so, Fostoria's designers created their colored novelties in the colonial-revival style, the dominant style of the 1920s. This choice was not unusual, for factory designers in the home-furnishings business knew that the mechanisms of popular culture, from mass magazines to art classes, had acclimated middle-class consumers to this visual mode. Just as important, designers understood contemporary rhetoric equating the colonial style with modernity and touting its adaptability to a range of decorating schemes. With all of this in mind, Fostoria's designers created lines such as Minuet, a topaz dinner service featuring delicate leaf-and-scroll motifs and shapes reminiscent of colonial glassware, "for the modern hostess" who appreciated "all the grace and charm of early America." Consumers had definite ideas about what constituted desirable glassware, and they were not easily persuaded to accept new taste preferences.[10]

To publicize its rainbow line, Fostoria engaged another group of fashion mediators, hiring one of the nation's leading advertising agencies, N. W. Ayer and Son, to create colorful ads for mass-market magazines such as *Vogue,* the *Delineator,* the *New Yorker,* and *Good Housekeeping.* Ayer's executives drew on the cultural discourse of the 1920s—on the rhetoric of glamour and gentility—in an effort to turn the idea of crystal upside down. In their promotions crystal became an old-fashioned product and Fostoria's pastel glassware was its modern, up-to-date replacement. In part Fostoria's advertising strategy entailed enlarging the portfolio of meanings ascribed to glassware so as to build brand-name recognition among women of different age groups. Some advertisements emphasized generational similarities in taste and behavior so that colored glass would appeal to grandmothers and youngsters alike. Others redefined dining traditions, presenting forty-two-piece luncheon sets as perfect accessories for the era of the servant shortage, easy eating, and the "dining-room-less house."[11]

During the Great Depression Fostoria's managers, like other American manufacturing executives, were troubled by the collapse of the U.S. business system. During the brief recovery at mid-decade, glassworks executives scrambled to strengthen their products' appeal among budget-conscious consumers, especially among those seeking to purchase stemware following the repeal of prohibition. China and glass buyers working for major retailers reported that the fad for chromatic etched glass tableware had run its course in the department-store trade. Middle-class consumers who reduced leisure expenses by entertaining at home were again buying crystal, choosing glassware that promised to sparkle in dimly lighted rooms. Ironically, Fostoria's calculated efforts to position chromatic glassware in middlebrow markets had backfired, for colored dinner services made by quantity-production factories in Ohio and Indiana had become the favored stock of five-and-ten-cent stores. Slowly but surely Fostoria's managers realized that something was wrong with the cogs of their product-development machinery, for designers and ad men had failed to anticipate or recognize these critical changes in demand.

As a result, Fostoria's managers, like others in the pottery and glass business, increasingly looked to professional buyers at retail stores as their primary intermediaries. Laboring at the point of purchase, these men and women earned their livelihood by scrutinizing consumers' preferences, discriminating between ephemeral fads and seminal shifts in taste and using these data to stock their stores with products that would sell.[12] Fostoria's managers reassessed their design and development practices, taking buyers' observations very seriously. A new generation of executives, C. B. Coe and William F. Dalzell, reduced the firm's output of tinted dinner sets and expanded its production of crystal stemware and decorative accessories, reshaping the composition of Fostoria's line to include colorless glassware that complemented the chromatic pottery tableware gaining popularity during the Depression. Fostoria's new managers also revamped the company's approach to markets, venturing into the five-and-ten-cent trade with inexpensive variations of their tinted glassware. Finally, they expropriated motifs from the fantasy world of drama for use in their crystal advertising campaigns. To be sure, they continued to use modernity and glamour—themes associated with Fostoria glassware in consumers' minds by the 1930s—but they recast those familiar subjects in a promotional campaign that drew on theatrical leitmotifs past and present.

Fostoria's Glass of Fashion promotion, dating from 1936, grafted

6.1. Glass of Fashion window or counter display, 1936, Fostoria Glass Company, Moundsville, West Virginia. Courtesy the Archives–Library Division, Ohio Historical Society, Columbus (OHS# 10485), and the Lancaster Colony Corp., Columbus.

Shakespearean tragedy onto Hollywood glitter, fusing histrionic motifs from one of the great texts of Western literature and images from one of the twentieth century's mass media, motion pictures. For the promotional campaign's title Fostoria's managers borrowed the euphonious, if tragic phrase of Shakespeare's Ophelia, who spoke of a mad Hamlet as formerly "the glass of fashion and the mould of form," that is, as a princely mirror of style and a model of deportment once admired by all at court. Ophelia's phrase was given a new twist when Fostoria crystal displaced Hamlet as the glass of fashion in magazine ads and promotional leaflets.[13]

Fostoria's executives and advertising consultants gave the Glass of Fashion promotion three-dimensional form in cardboard display units for jewelry stores, china shops, and department stores. Each window or countertop unit depicted two glamorous Hollywood actresses on a grand set and featured space for pieces of Fostoria glass. Writing to storekeepers, the firm's sales manager boasted that these dramatic props would ignite women's desires to buy Fostoria glassware. Indeed, the careful juxtaposing of glittery Hollywood images with sparkling glassware would lead consumers to associate wealth, status, and glamour with Fostoria. The Glass of Fashion display suggested that Fostoria's products were capable of injecting consumers' mundane lives with a touch of upper-class cinematic elegance (fig. 6.1).[14]

Fostoria's marriage of old and new in the Glass of Fashion campaign—the fusion of Shakespearean text and Hollywood imagery—was designed to overcome generational tensions, if not simply differences in perception, exacerbated during the trying economic times of the Depression. The reference to Hamlet was undoubtedly recognizable to college-educated, middle-class consumers, but it was especially resonant among great-grandmothers, who had come of age in the nineteenth century, when Shakespeare's plays had enjoyed popularity in the United States. In contrast, Fostoria's cinematic allusions strongly reverberated among younger women, whose favorite entertainments included motion pictures. Whether they admired independent-minded stars such as Katherine Hepburn or femmes fatales such as Marlene Dietrich, film audiences partook in a world of luxury and success on the screen. And they might experience a modicum of elegant living, or own a pleasing fragment of it, by purchasing and using Fostoria crystal.[15]

During the 1930s Fostoria glassware was indeed a mirror of fashion, a model of manners, and more. Without question, the firm's advertisements reflected the propensity of Depression era shoppers to seek psychological relief from larger social and economic exigencies through the consumption of dazzling, uplifting, and modestly priced consumer products. But Fostoria's managers also offered consumers a model of deportment that venerated elegance and formality, that is, they touted nineteenth-century gentility disguised in modernistic clothing. During the Depression, as before, membership in the right class—the middle class—was a slippery thing that people might acquire through the accumulation, display, and use of the right stuff. Fostoria's managers were offering consumers a new cinematic variation on the very old Cinderella theme.

Significant demographic shifts following World War II provided American manufacturers of furniture and household accessories, including Fostoria, with tremendous opportunities for increased sales and profits. Between 1940 and 1960 the marriage rate increased by 8 percent; 60

percent of the population was married in 1940, 66 percent in 1950, and 68 percent in 1960. Another salient demographic characteristic of these decades was the young age of newlyweds. From the late 1940s through the 1960s the median age at first marriage for women was around twenty; for men it was around twenty-two. As the marriage rate among young adults (and older ones) soared, so did sales of the commonplace necessities of first-time housekeeping, from dinner plates and sherbet dishes to casseroles and lemonade tumblers. As more couples wed, experts in manufacturing and merchandising were charged with making new products and advising homemakers on the use of these household goods.[16]

Once again Fostoria's executives rode on the crest of cultural change, skillfully negotiating the slippery marketplace where gender ruled absolute. In the postwar period Fostoria's managers targeted brides and brides-to-be as a distinctive audience, that is, as a group with special needs in terms of products, advertisements, and publicity campaigns. Although gifts of china, glass, and silver had long been associated with weddings and marriage, the so-called bridal market as we know it today, replete with registries, is relatively new. A creation of *Bride's* editors, registries dated from the early 1930s, but shortages of men and materials temporarily slowed the marriage rate and dampened the trade in wedding gifts during World War II. After the war, journalists, advice-book authors, retailers, and manufacturers again encouraged engaged women to acquire household necessities by promoting hope chests, bridal registries, and showers. By 1953, depending on their economic class, brides-to-be were being given from one to six showers by their families, friends, or coworkers. Reminiscent of courtly balls, showers and parties dedicated to gift acquisition became part and parcel of marriage rituals as aspiring Cinderellas constructed fantasies of the good life and celebrated promising futures by accumulating household accessories, including china and glassware.[17]

In the postwar era managers and public-relations consultants working for Lenox China, a New Jersey pottery that became the nation's largest supplier of porcelain tableware to brides, aggressively promoted the bridal gift registry system. Fostoria followed on Lenox's coattails to became the country's leading manufacturer of glassware for brides. The principle behind china and glass registries was simple. Before her wedding shower a woman visited a retailer, studied china and glass patterns, chose her favorites, and listed her preferences with the store's bridal secretary. The bride-to-be then informed her friends and family of her listing, or registration, with a particular retailer. Those people, in turn, could purchase place settings on behalf of the bride. Under this merchandising system an engaged woman could acquire extensive services of sometimes costly china and glass. Certainly bridal registries streamlined the distribution of china and glassware through the marketplace, increasing sales and bringing profits to manufacturers and merchants. But registries also had significant cultural functions. The selection of shapes, patterns, and forms was almost endless, and a bride could specify a distinctive combination of china and glassware within almost any price range. This relatively unique ensemble, reflecting the individuality and taste of the bride-to-be, was the modern equivalent of Cinderella's form-fitting slipper. Perhaps more important, the registry system offered easy gift solutions to the bride's older relatives and friends who still believed in the practice of formal dining. In this way bridal registries helped to bridge the generation gap that Fostoria's managers had struggled to overcome with product design and advertising in the 1920s and 1930s.[18]

By 1946 Fostoria was already the glassware of choice among teenage compilers of hope chests, and the company's managers sought to augment their sales to young brides-to-be with another innovation in public relations. Once the herald of modernity, Fostoria embraced a legacy of craftsmanship during the late 1940s and 1950s. At that time Fostoria's sales staff used images and text to link their firm and its female customers to a bygone era, anchoring American women to the culturally sanctioned roles of housekeeper, handmaiden, and hostess. American culture had grown more conservative, with marriage, family, and home life assuming a new significance. Fostoria, the glass chameleon, adapted to those changes with updated versions of the Cinderella story.[19]

To carve a market niche in the postwar environment Fostoria's executives again drew on the expertise of fashion intermediaries, those critical experts on taste who had served the firm in earlier decades. To be sure, Fostoria's managers had learned an important lesson from retail buyers during the 1930s. Fostoria's new consumer liaisons differed from their counterparts of the Jazz Age in that they often were women laboring close to consumers rather than men working as product designers or ad executives.

The gender of this new generation of fashion experts was an asset to manufacturers; few would disagree with the opinion that women could best decipher the mysteries of the gendered marketplace. Whether they worked as buyers of china and glass for department stores or bridal gift secretaries in china shops, merchandisers working at the point of sale had their fingers on the pulse of consumer taste. Women working in the retail trade watched and listened as customers studied, touched, and discussed goods on the selling floor. They then formulated profiles of distinctive consumer groups in their minds and drew on these profiles when restocking their departments, collaborating with manufacturing executives on new products, and helping factory sales managers to develop special promotions. Again and again Fostoria's managers depended on the expertise of these fashion intermediaries as they created new products and promotions.

Among Fostoria's key links to the marketplace was Alice Bayse, the bridal gift secretary at J. L. Hudson Company, a department store in Detroit, who routinely corresponded with the firm's sales manager, David B. Dalzell. In a letter to Dalzell, Bayse praised the layout of Fostoria's new promotional booklet for glassware for brides-to-be. Bayse said that most of the women visiting her department were indeed concerned with coordinated table appointments, correct seating arrangements, and the difference between modernistic and traditional styles. But, she confided to Dalzell, few of her customers really understood the grammar of ornament, rules of etiquette, or the advantages of open-stock patterns. Believing that manufacturers and merchandisers were responsible for educating consumers as well as catering to their desires, Bayse suggested that Dalzell modify his publication so as to enhance consumers' comprehension of Fostoria's line of formal dining accessories. Indeed, a glassworks executive like Dalzell could not ignore Bayse, for this Detroit merchandising expert was one of his critical lifelines to consumers. Her feedback on style, fashion, and gender was invaluable.[20]

Several of Fostoria's postwar promotions also demonstrate the central role of other types of fashion intermediaries in reading the nuances of female taste preferences, which ultimately governed choices at the sales counter. In 1945 Fostoria's executives unveiled their ambitions for expanding sales among young Cinderellas, or brides-to-be, when they instituted a national promotion aimed at high-school classes in home economics.

Save this picture for future identification

In thousands of high schools and colleges, Home Economics teachers are using Fostoria literature to tell their pupils how to choose and use glassware. In their notebooks the girls have a Fostoria folder as their text on the subject.

You are likely to meet these girls someday, as brides, in your China and Glass department. They will want you to show them good crystal. And they will know that Fostoria *means* good crystal. That's when you will cash in on this Fostoria advertising that is directed at the schoolgirls of today, the brides of tomorrow.

DON'T PASS UP THE OPPORTUNITY to make this Fostoria educational advertising work for you. Check to see that Home Economics teachers in your schools know about it. Show it to them and offer to co-operate in classroom projects. If you don't have samples of the literature, write for it.

FOSTORIA GLASS COMPANY
MOUNDSVILLE, WEST VA.

52 CROCKERY & GLASS JOURNAL for March, 1945

6.2. Fostoria Glass Company, "Save This Picture for Future Identification," *Crockery and Glass Journal* 136 (Mar. 1945): 52. Courtesy Lancaster Colony Corp., Columbus.

The mediators in this project were two professors of home economics. With the assistance of these consultant consumer liaisons, Fostoria's managers developed and distributed instructor's manuals, classroom charts, textbooks, and films explaining the history, selection, care, and use of handmade glassware to teenage girls (fig. 6.2).

The centerpiece of Fostoria's home-economics program was a film, *Crystal Clear,* which claimed that well-appointed tables set with handmade Fostoria glassware were "as spiritually important" to building middle-class, family-oriented values as was "good food to our health." But *Crystal Clear* did more than convey the message that fine dining was the adhesive that would hold postwar families together. The film also entwined images of craftsmanship and domesticity to disassociate male glassblowers and female consumers from the "twentieth-century world" of production and tie them to a premodern fairy-tale world of tradition. The movie's glassblower was endowed with craft skills linking him to the past; he depended "on the touch of his own hands and the control of his own breath" rather than on the power of "compressed air and conveyor lines" to create objects of "lasting beauty." Fostoria's artisan symbolized a vanished era, a period when a craftsman's skills were emblematic of his masculinity; by the postwar years these skills had been marginalized in most industries, although they were still vibrant in Fostoria's factory. The film's female protagonist, the teenage Mary Lee, tries to ensnare "the boy next door" by demonstrating her thorough understanding of handmade glass, knowledge gained in her home-economics class. Mary Lee explains glassmaking technology and marketing to her potential mate, progressing from the glass furnace to the department store. The film's nubile Mary Lee was guardian of two types of know-how, handicraft and consumption. Both spheres of knowledge were excluded from the spheres of contemporary middle-class and working-class men, who labored in offices or on assembly lines. In the context of the film Fostoria was no longer the glass of fashion, the stuff of Hollywood glitter and glamour. Fostoria was now the glass of tradition, a conservative product that cultural experts had deemed appropriate for inexperienced young brides, for postwar Cinderellas awaiting the arrival of white- or blue-collar princes.[21]

Nowhere was Fostoria's new theme of tradition better expressed than in the firm's bridal display, designed in 1949. Whether they shopped at the

6.3. Bridal window or counter display, 1949, Fostoria Glass Company, Moundsville, West Virginia. Courtesy of the Archives–Library Division, Ohio Historical Society, Columbus (OHS# 10484), and the Lancaster Colony Corp., Columbus.

Model Grocery Company in Pasadena, California, or at the Hess Brothers' department store in Rockford, Illinois, women encountered Fostoria's image of ideal womanhood in retail china-and-glass departments across the country. Like *Crystal Clear,* Fostoria's bridal display evoked a sense of timelessness by conflating images from various eras, by fusing iconography of the past, present, and future. Fostoria's cardboard bride had a modern hairstyle, but she was garbed in a ruffled white dress suggestive of tight-laced fashions from the nineteenth century. She had a demure manner and the slender, yet buxom figure of a calendar girl; her eyes avoid onlookers as she admired the Fostoria glassware at her feet. She stood next to oversized wedding invitations with texts suggesting that Fostoria glassware was suitable for brides, newlyweds, and wives. The bridal display thus invited all women to buy Fostoria's timeless goods. The display and use of these goods in a home promised to preserve a woman's youth, beauty, and allure, the ultimate goal of postwar Cinderellas (fig. 6.3).[22]

While firms such as Fostoria targeted brides-to-be in the postwar years, other manufacturers catered to the burgeoning market for home furnishings in very different ways. Managers in larger, quantity-production glassmaking firms such as the Libbey Glass Division of the Owens-Illinois Glass Company focused not on the market for fancy dishes but on the growing demand for casual household accessories. By midcentury the cult of gentility had lost much of its vitality, as indicated by the tremendous efforts on the part of firms like Fostoria and Lenox to build new audiences among extremely young women. Concurrently, American vernacular culture reared its head as people insisted on compressing the most time-intensive elements of consumption into streamlined forms. Vernacular modes of behavior were grafted onto highbrow and middlebrow customs, so that the new cultural modus operandi was a hybrid of formality and informality, a crossbreed best described as casual.[23]

Casual living emerged as an alternative American cultural style partially in reaction to the high modernity of the interwar and war years, that is, as a tension-releasing response to the recent roller-coaster ride of boom, bust, and boom. But easy living also coincided with the start of a major population shift away from the modern metropolis of concrete and steel toward greener suburbs. To complicate matters, informality was by no means new; working-class people, farmers, immigrants, racial minorities, and impoverished Cinderellas had long eaten in the kitchen, so to speak. In fact the cultural convention of easy eating had germinated during the Great Depression and reared its head during the demographic shifts of the postwar years. Just as gentility made Americans conscious of their manners and their table accessories, easy living enforced rules for behavior and defined parameters for the use of consumer goods, including glassware.

As always, Fostoria's managers read the marketplace and created glassware geared toward changing tastes, grafting the rhetoric and iconography of casual living onto advertisements and public-relations campaigns for postwar brides. But casual living allowed consumers far greater expression than had gentility, and women happily engaged this freedom of material choice. Female shoppers were a discerning lot whose tastes were diverse but not mutable. The beauty of a piece of glassware depended on the perceptions of the viewer, who brought a lifetime of visual and tactile experiences to reading the aesthetics of simple objects, whether vases, wine glasses, or sherbet dishes. An artifact that was pleasing in the eyes of a Manhattan socialite may have appeared ostentatious to a farm girl from Arkansas or gaudy to an immigrant factory worker in Illinois. The number of interpretations, the number of uses for a single product, was seemingly endless. Some women appreciated china and glassware as shiny, colorful ornaments. Socially ambitious women associated glass goods with enviable upper-class lifestyles. Still others wanted utilitarian kitchen utensils that were sanitary and simple to use. The postwar market was highly variable, changeable, and difficult to master. Some women continued to embrace the Cinderella paradigm, engaging objects as mechanisms for creating comfortable domestic sanctums. But others rejected the Cinderella model, seeking fulfillment beyond the private sphere of the home and its assemblage of artifacts. Even within this shifting framework glass objects still functioned as meaningful repositories for many female consumers.

Nothing speaks to continuity and change in the Cinderella paradigm more than the experiences of individual consumers. First-person testimonies about the significance of objects in daily life are rare, as few people consciously engage their artifactual vocabularies to write about specific material experiences. In the absence of written testimony from consumers, historians can draw on data from market surveys to explore the contours of people's tastes and consumption patterns. In this vein, a survey of newlyweds conducted by editors at *McCall's* during the early 1950s provides insight into consumers' material experiences and perceptions in the postwar era, constituting a fitting ending to our series of Cinderella stories.

Most of the Cinderellas in the *McCall's* bridal survey were proud of their accumulations of silver, china, and glassware. Among one Vermonter's wedding gifts were goblets, sherbet dishes, candlesticks, and a fruit bowl in her chosen glass pattern; additional presents included Pyrex casseroles and roasters. This farm woman rarely entertained in the evenings, so her good chinaware sat in the cupboard until Sunday, when she always used "her good dishes and good silver" for a big meal. Upon occasion parents or friends attended these fetes, but often food was shared only by the couple. Wealthier suburban brides also enjoyed owning fancy china and glass. Betty, a twenty-six-year-old journalist, had married a fellow news reporter, and the couple had settled in a suburb of Atlanta. Among her 172 wedding gifts were china, crystal, "heirloom cut glass," "an earthenware everyday china set," casseroles and chafing dishes, an "antique decanter and glasses in a silver holder," and a "modern Swedish crystal glass decanter." A combination of old and new, the Atlanta bride's gifts would have been the envy of any modern Cinderella.[24]

Although these women, their families, and their friends agreed that china and glass were household necessities, many of them diverged regarding what constituted the proper use of these goods. Some women emulated the models put forth by manufacturers, magazine writers, and other culture brokers when using household accessories. They hosted evening cocktail parties or afternoon bridge games, at which they used glassware to serve beverages. When the *McCall's* interviewer visited Betty's house near Atlanta, the journalist demonstrated her awareness of cultural prescriptions by serving the interviewer lunch on "wedding-present china and silver." Betty reported that she entertained friends "two or three times a week with informal cocktail parties, dinner parties, and very casual late suppers"; she also intended to "give some bridge parties after the holidays." In contrast, once again, was the rural Vermont bride. Although she may not have cared about cultural prescriptions, this woman clearly believed that her china and glassware were ritualistic objects. Using her finery for solitary Sunday dinners with her husband was a silent tribute to the success of their marriage. In this respect her china and glassware became, as Fostoria's executives had hoped, anchors of experience and emblems of femininity.[25]

Some women discarded cultural prescriptions for the use of china and glassware, challenging the happy-go-lucky depictions of domesticity in postwar magazines, television shows, and movies. Some recently married couples were clearly less concerned with material comforts than their parents or grandparents. One bride's mother happily reminisced about the early years of her marriage, fondly recalling the challenge of creating a comfortable home on a tight budget. "When I married," she noted, "I wanted a home and security" as embodied in "my own home and own furniture." Much to this mother's dismay, her daughter and son-in-law were indifferent to the material world. Oblivious to their surroundings, the newlyweds were satisfied with a "dreary one-room furnished apartment" as long as they were together. Like most of the one hundred newly married women in the *McCall's* survey, this bride worked, and she had little time to devote to homemaking. Many working couples dwelled in sparse, sometimes dingy apartments furnished with "cast-offs from their families." Many a working woman believed that her slice of the American pie would come her way after her husband graduated from college, obtained a new job, or was better established in the community.[26]

The pretty picture of suburban living that appeared in postwar advertisements was also challenged by the testimony of women who owned the appropriate household accessories. Some brides of the 1950s clearly had greater expectations than their grandmothers; education and ambition led some newly married women to question the value of possessions. One college graduate in Oakland, California, received matching tableware and crystal as wedding gifts, but she rarely used these goods. Trapped in an unhappy marriage, this bride viewed her china and glass as symbols of her subordination to a domineering, jealous husband who forbade her to entertain friends in their home. Clearly, marital bliss was no longer assured by the acquisition of the right stuff as promised by the Cinderella tale. But as illustrated by this final example from *McCall's,* women still used glassware as repositories of meaning, even if those meanings were unpleasant and dark.[27]

Although the original Cinderella story had lost some of its persuasiveness as a model for male-female relationships by the mid–twentieth century, this romance still resonated in American culture in other important ways. Like other fairy tales, the Cinderella story betokened withered and wishful worlds, that is, a long-gone society wherein social rank and inheritance went hand in hand and a fantasy land wherein emotional fulfillment and financial success were gift-wrapped in prince's clothing. But more germane, the Cinderella romance serves as a reminder that we live in a material world, that our things—although far less glamorous than Cinderella's regal possessions—function as talismans in intimate relationships, as symbols of personal hopes and dreams, and as moorings for cultural beliefs and values. Indeed, few people come of age without learning how to use artifacts to mediate social circumstances, to construct self-identity, and to help situate individuals and groups within the culture. From childhood we learn about the power of things in diverse ways, by observing the roles of artifacts in grownup dramas or by reading fairy tales about objects' magical powers. But people-object interactions are so common in our culture that we often overlook their significance, both in daily life and in our study of the past. Fantasies like the Cinderella story are historical hooks that engage our attention and analytical tools that help us to think critically about the material world and its commonplace goods, including glassware.

Few understood the crucial role of ordinary objects—and people's fantasies about those objects—in the construction of self- and cultural identity in nineteenth- and early-twentieth-century America better than the manufacturing executives directly responsible for giving form to consumers' material wishes. Certainly no others were more aware of women's desires, for the wants of female customers motivated decision makers in the home-furnishings trade. In this gendered marketplace producers yielded to their audiences in cultural production, and glass manufacturers were no exception. Working through the channels of fashion intermediaries, glassmakers gathered data about female fantasies of the good life and translated those dreams into reality as beautiful, fragile glass objects. At Fostoria consumer demand drove corporate strategy, rather than vice versa, as managers carefully labored to meet even the most fantastical expectations of would-be Cinderellas.

Material circumstances constrained the flow of information among, as well as the actions of, glassworks managers, fashion intermediaries, and consumers, for technology, aesthetics, and economics always circumscribe cultural production. Unlike Cinderella's godmother, Fostoria's managers felt the bridle of their material circumstances, for they did not own magic wands. Nonetheless, Fostoria's managers—along with others in the glass business—created saleable products appreciated by generations of women. Cognizant of the inherent human need to infuse possessions with meaning, Fostoria's executives orchestrated the production of artifacts in timeless styles that matched object ensembles in any number of decorating schemes, formal or casual. The firm's flexible production methods enabled managers to pursue a mixed-output strategy—making glassware for myriad tastes and pocketbooks—well into the mid-twentieth century. Women sometimes ignored cultural prescriptions and adapted glass objects to fit individual circumstances. In doing so, consumers behaved exactly as anticipated by Fostoria's managers, creators of the glass of fashion. Like fairy-tale princesses, consumers acquired and used glassware as repositories of meaning, as anchors of memory, and as symbols of fantasy. Such was the legacy of the Cinderella story in the gendered marketplace of modern America.

Notes

I thank Anne M. Boylan, Roger Horowitz, Jacqueline McGlade, Arwen Mohun, Bruce J. Schulman, and others who provided constructive criticism on earlier drafts of this chapter, presented at the Ninth Berkshire Conference on the History of Women, Vassar College, Poughkeepsie NY, in July 1993; at the "His and Hers" conference in April 1994; during Women's History Month at Monmouth University, Monmouth NJ, in March 1996; and at the Americanists' Forum, Boston University, in April 1996.

1. Countless versions of "Cinderella" have been published in the West over the centuries. This chapter draws on Joanna Cole, *Best-Loved Folktales of the World* (Garden City NY: Doubleday, 1982), 3–8; and Wanda Gág, *Tales from Grimm* (New York: Coward, McCann & Geoghegan, 1936), 99–120. For a recent analysis of the tale that focuses on envy see Melanie Thernstrom, "The Glass Slipper," in *Women on Ice: Feminist Essays on the Tonya Harding/Nancy Kerrigan Spectacle,* ed. Cynthia Baughman (New York: Routledge, 1995), 148–61.

2. Unless otherwise noted, this discussion of the production, distribution, use, and meaning of pottery and glassware is drawn from the following works by Regina Lee Blaszczyk: "Imagining Consumers: Manufacturers and Markets in Ceramics and Glass, 1865–1965" (Ph.D. diss., University of Delaware, 1995); "The Aesthetic Moment: China Decorating, Consumer Demand, and Technological Change in the American Pottery Industry, 1865–1900," *Winterthur Portfolio: A Journal of American Material Culture* 29 (summer/fall 1994): 121–53; "'Reign of the Robots': The Homer Laughlin China Company and Flexible Mass Production," *Technology and Culture* 34 (Oct. 1995): 152–200; "Imagining Consumers: Manufacturers and Markets in Ceramics and Glass, 1865–1965," *Business and Economic History* 25 (fall 1996): 13–18;

and *Imagining Consumers: The Business of Product Design and Innovation* (Baltimore: Johns Hopkins Univ. Press, forthcoming).

3. For a fuller discussion of fashion intermediaries see Blaszczyk, "Imagining Consumers"; and Grant McCracken, *Culture and Consumption: New Approaches to the Symbolic Character of Consumer Goods and Activities* (Bloomington: Indiana Univ. Press, 1990), ch. 5.

4. On the company's history see "Fostoria Reveals the Magic in Glass," *West Virginia Review* 14 (Nov. 1936): 74–75. On flexible production see Philip Scranton, *Proprietary Capitalism: The Textile Manufacture at Philadelphia, 1800–1885* (New York: Cambridge Univ. Press, 1983); and idem, *Figured Tapestry: Production, Markets, and Power in Philadelphia Textiles, 1885–1941* (New York: Cambridge Univ. Press, 1989).

5. On formal dining see Susan R. Williams, *Savory Suppers and Fashionable Feasts: Dining in Victorian America* (New York: Pantheon, 1985); and Richard L. Bushman, *The Refinement of America: Persons, Houses, Cities* (New York: Alfred A. Knopf, 1992), ch. 12.

6. On servants see Glenna Matthews, *"Just a Housewife": The Rise and Fall of Domesticity in America* (New York: Oxford Univ. Press, 1987), ch. 4; and Ruth Schwartz Cowan, *More Work for Mother: The Ironies of Household Technology from the Open Hearth to the Microwave* (New York: Basic Books, 1983), chs. 1, 5. On dining habits see Harvey A. Levenstein, *Revolution at the Table: The Transformation of the American Diet* (New York: Oxford Univ. Press, 1988), chs. 12–14.

7. On the symbolic meaning of consumer goods see the essays in Arjun Appadurai, ed., *The Social Life of Things: Commodities in Cultural Perspective* (New York: Cambridge Univ. Press, 1986); and Leigh Eric Schmidt, *Consumer Rites: The Buying and Selling of American Holidays* (Princeton: Princeton Univ. Press, 1995).

8. For a description of these new lines see *The New Little Book about Glassware* (Moundsville WV: Fostoria Glass Company, 1928), box 7, Fostoria Glass Company Records (hereafter FR), Archives–Library Division, Ohio Historical Society, Columbus, Ohio.

9. On the vogue for color see Rexford Daniels, "The Color Wave Here to Stay," *Pottery, Glass and Brass Salesman* 37 (26 Apr. 1928): 17, 23, 26; Herbert C. Hall, "Color—Is It a Fad?" *Magazine of Business* 54 (Sept. 1928): 245–82; E. H. Brown, "The Paint Brush Knocks at the Sanctum Door," ibid., Dec. 1928, 655; and "Color in Industry," *Fortune* 1 (Feb. 1930): 85–94. On color in the kitchen see Genevieve A. Callahan, "Bringing Color to Your Kitchen," *Better Homes and Gardens* (hereafter *BHG*) 3 (June 1925): 16, 52; "The Movement for Colored Kitchen Utensils Is Spreading," *Pottery, Glass and Brass Salesman* 36 (6 Oct. 1927): 15; "Colored Kitchen Wares," ibid., 15 Dec. 1927, 167; "More and More Colored Wares," ibid., 24 Nov. 1927,

33; Armstrong Cork Company, "Color Is the Cure for That Tired-Looking Kitchen!" advertisement in *BHG* 6 (Apr. 1928): 95; George D. Roper Corporation, "The Climax of Kitchen Color Harmony," advertisement in ibid., May 1928, 92; and Standard Sanitary Manufacturing Company, "Let This Radiant Sink Gladden Every Busy Hour," advertisement in ibid. 7 (Feb. 1929): 2.

10. Fostoria Glass Company, "The Exquisite Minuet," n.d. (quotations), and "Mayfair," n.d., both in box 7, folder: Old Leaflets, FR; "Modernistic Note Featured in Fostoria Designs for 1929," *Crockery and Glass Journal* (hereafter *CGJ*) 107 (Jan. 1929): 48; Alden Welles, "Fostoria Glass: A Pioneer in the Field and a Standard Bearer in a Forward-Marching Industry," ibid., Sept. 1929, 24–27, 81.

11. For the quotation see Hazel T. Becker, "Four Dining-Room-Less Homes," *BHG* 4 (June 1926): 14–15. See also Fostoria Glass Company, "Four Years Ago We Began to Advertise Fostoria," advertisement in *CGJ* 106 (5 Jan. 1928): 7; and Advertising Proof Book 266: Fostoria Glass Company, 1924–1928, N. W. Ayer and Son Collection, Archives Center, National Museum of American History, Smithsonian Institution, Washington DC. On the context for Ayer's approach see Roland Marchand, *Advertising the American Dream: Making Way for Modernity, 1920–1940* (Berkeley: Univ. of California Press, 1985), ch. 5.

12. J. Walter Thompson Company, "Corning Glass Works, Interviews with Department Stores on Glassware," Aug. 1937, 6, 8, 10–11, 14–15, 20, 22, in Microfilm Collection (16 mm.), reel 38, J. Walter Thompson Archives, Special Collections Department, William R. Perkins Library, Duke University, Durham NC. On the critical role of retail buyers in product innovation see Blaszczyk, "Imagining Consumers," chs. 3, 5; and Joseph M. Wells, "Buyers' Ideas Create 90 Percent of New American Dinnerware," *CGJ* 114 (Feb. 1934): 25, 38.

13. Shakespeare, *Hamlet* 3.1.153, in *William Shakespeare: The Complete Works*, ed. Alfred Harbage (New York: Viking, 1969), 951.

14. "Fostoria: The Glass of Fashion," consumer leaflet, n.d., box 7; "A Distinctive Display Unit for Your Counter or Small Window," photograph X9344, illustrating dealer's display unit, n.d., box 3; and photograph AZ444, [1936], box 3, folder: Advertising, 1936—Fostoria Display, all in FR.

15. Lawrence W. Levine, *Highbrow/Lowbrow: The Emergence of Cultural Hierarchy in America* (Cambridge: Harvard Univ. Press, 1988), ch. 1.

16. Elaine Tyler May, *Homeward Bound: American Families in the Cold War Era* (New York: Basic Books, 1988), 6, 20.

17. On hope chests see Lee M. Chapman, "The Hope Chest: A Service for Brides," *Good Housekeeping* 130 (Jan. 1950): 26–27; James Mayabb, "The Hope Chest: A Service for Brides," ibid., Feb. 1950, 26–27; and idem, "The Hope Chest: A Service for Brides," ibid., Apr. 1950, 26–27. On showers see Lucy Goldthwaite et al., "At

Home with Young American Brides," ca. 1953, 10–11, Russel Wright Papers, ser. 4, box 1, folder: American Brides, *McCall's,* Arents Library, Syracuse University, Syracuse NY.

18. On Lenox see Ellen Paul Denker, *Lenox China: Celebrating a Century of Quality, 1889–1989* (Trenton: Lenox and New Jersey State Museum, 1989), 70; John Tassie, interview by author, 24 Sept. 1992, Princeton NJ; and Robert Sullivan, interview by author, 15 Oct. 1992, Princeton NJ. Tassie and Sullivan were top executives at Lenox in the postwar era; tape recordings of both interviews are in the possession of the author. A place setting consisted of all the china or glass used by one person in a formal meal; it might include a dinner plate, a salad plate, a bread plate, a dessert plate, a fruit bowl, and a cup and saucer.

19. David B. Dalzell, Factory Correspondence (hereafter FC) 217, 14 Nov. 1946, box 5, folder 10: Sales Letters, 1946–1947, FR.

20. Alice Bayse to Dalzell, 30 Nov. 1949, box 2, folder 3, ibid.

21. For the quotations see Helen Hunscher and Blanche Harvey, *Crystal Clear: A Teacher's Guide* (Moundsville WV: Fostoria Glass Company, 1947), p. 7, box 5, folder 9: Salesmen Letters, 1947, FR. Hunscher and Harvey were professors of home economics at Western Reserve University in Cleveland, Ohio. On Fostoria's home-economics program see Merlin DuBois, FC 182, 9 Dec. 1944, box 5, folder 6: Sales Letters, 1944; DuBois, FC 186, 26 Mar. 1945, box 5, folder 7: Sales Letters, 1945; DuBois, FC 204, 27 Mar. 1946, box 5, folder 8: Sales Letters, 1946; Dalzell, FC 233, 5 Mar. 1947, box 5, folder 10: Sales Letters, 1946–1947; and DuBois, FC 238, 22 Nov. 1947, box 5, folder 9: Salesman Letters, 1947, all in FR. On home economists and business see Regina Lee Blaszczyk, "'Where Mrs. Homemaker Is Never Forgotten': Lucy Maltby and Home Economics at Corning Glass Works," in *Rethinking Home Economics: Women and the History of a Profession,* ed. Sarah Stage and Virginia B. Vincenti (Ithaca: Cornell Univ. Press, 1997), 163–80; Carolyn Manning Goldstein, "Mediating Consumption: Home Economics and American Consumers, 1900–1940" (Ph.D. diss., University of Delaware, 1994); and idem, "From Service to Sales: Home Economics in Light and Power, 1920–1940," *Technology and Culture* 38 (Jan. 1997): 121–52.

22. On the bridal display see Dalzell, Moundsville WV, to Norval Slater, Chicago; "Point-of-Sale Display for Bridal Promotion"; "Partial List of Dealers Shipped Bridal Display"; and "Fostoria Helps You Step Up Bridal Sales", all from the year 1949 and located in box 2, folder 11, FR. Dalzell encouraged Slater, Fostoria's representative at the Merchandise Mart in Chicago, one of the nation's largest home-furnishings showrooms, to "push" the bridal display to visiting retail buyers. Slater in turn urged retailers to use the displays in conjunction with newspaper advertising campaigns created from illustrations provided by the glassworks.

23. For more on easy living, pottery, and glassware see Regina Lee Blaszczyk, "The Wright Way for Glass: Russel Wright and the Business of Industrial Design," *Acorn* 4 (1993): 2–22.

24. Goldthwaite et al., "At Home with Young American Brides," 35, 44–53.

25. Ibid., 44–53.

26. Ibid., 6.

27. Ibid., 58–65.

Seven

Shopping for a Good Stove

A Parable about Gender, Design, and the Market

Joy Parr

We are accustomed to thinking of women as shoppers and shoppers as women. "Born to shop" bumper stickers are affixed (whether sardonically, wearily, or proudly) to cars driven by women and not to those driven by men. Described for a decade as something women do by nature, that they are born to, shopping lately has been proclaimed by talk-show hosts and writers of mass-market paperbacks a women's addiction, a disease to which women particularly are prone.

There was, of course, a time when no one, whether female or male, shopped—when necessaries were found or chased or made or traded—and thereafter a time when men turned their family's produce or remittances into cash and then goods on their trips alone to town. Women in modern times have been made shoppers by their circumstances rather than by their nature. But these circumstances, the changing social setting and experience of shopping, have not been much studied, partly because thinking about shopping, in our time ambiguously work and weakness, at once invisible and obsessional, makes women scholars nervous.

Shopping is not always the same experience. This chapter is about shopping for a stove in the central Canadian province of Ontario in the years 1950–55, about trying to find a particular good to meet a clearly but variously defined need; a good made in a limited number of forms by a finite number of firms and sold according to studiously defined practices by well-established stores. The aim is to get to know women of that time better by studying their experience as shoppers: the assumptions about womanliness acted out in the shops and built into the goods for which women shopped; the ways in which women were forbidden but nevertheless found out what they needed to know in order to buy; and the latitude they had or lacked to get manufacturers to make for them what they wanted to use. This chapter is first about women in Ontario in the early 1950s and second an experiment in diagnosing the power politics of shopping. It is finally a parable about a stove.

After ten years of depression, six years of war, and five years of disruption in the return to peacetime production, a substantial portion of Ontario households were in need of a stove. In 1950 a stove, unlike a refrigerator, was an undisputed household necessity. Unlike the refrigerator, it was rarely touched by men, used almost exclusively by women to perform tasks most men had not mastered. Stoves were major purchases, but their principal users were not usually major income earners. They were manufactured in both Canadian firms and American branch plants according to drawings made by male designers and sold by male commission salesmen. Thus stoves provide a particularly good prism for investigating gender politics in the postwar world of goods.

Being Sold and Buying

I have found no systematic study of who shopped in postwar Ontario, but writers in the trade and business press urged their readers to recognize that most of their customers were women, to "cultivate the ladies, they do 90% of retail buying," and having acknowledged woman as the "nation's purchasing agent," to "slant . . . sales and merchandising programs to attract the feminine purchaser." To impress upon their staff that women were important buyers whose "whims" must be catered to, in 1950 the men in the Merchandise Office at Eaton's in Toronto cited 1948 Illinois shopping data showing that 41 percent of all electrical appliances were bought by women alone, and 21 percent by women and men together. The Illinois study suggested that women in lower-income groups and in larger urban centers did even more of the family buying. Still, from the appliance retailer's perspective there was at least "the shadow of a man behind every women who buys," and in one in five cases the ignoble presence of "an impatient husband who wants to see how his money is being spent but is easily bored by his wife's inability to make up her mind in a hurry."[1] Retailers assumed that women "purchasing agents" reported to male bosses.

In the early 1950s major appliances were sold by male commission salesmen. Female sales staff might have carried "a more convincing story to the customer." Both the home-economics director and the general sales manager at Moffat's, a leading Ontario stove manufacturer, recommended as much in 1952. But the prevailing view in the trade was that "saleswomen in the appliance field have fallen flat on their faces. Men don't like them on the selling floor. Women customers don't like to be sold by another woman. Sales girls often dress up too much. Load themselves with expensive hairdos and jewellery. They set up sparks when they came in contact with a poorly dressed housewife dragging two howling kids with her. Some women have succeeded but very few." We do not know what women of the time really thought about buying from appliance saleswomen or being appliance saleswomen. What is clear is that men dominated the well-paid jobs on appliance sales floors, and a shopper stepping onto that floor in search of a stove was entering a place in which masculine assertions and intentions defined womanhood.[2]

The postwar conventions of good salesmanship—"virile salesmanship," as it was called in a story about Moffat's—were defined man to man.[3] Adapting these conventions for men selling to women mattered to both manufacturers and retailers given the volume and intense competition in appliance sales in the early 1950s. But given prevailing gender roles, making work that required men to serve women palatable and effective was a brisk managerial challenge. It was easy enough to advise men, most comfortable thinking of their wares as machines, to adopt women's diction and speak in terms of cooking speeds, not wattage, of the number of baking dishes, rather than square inches, an oven would hold.[4] But consider the stalwart principles of the salesman's creed: "Sell yourself, you are part of the package your prospect buys"; "Establish some common interest with your buyer"; "Be a good listener"; "Arouse your prospect's interest." These took on a salacious edge when a salesman was making a pitch to a woman. Cecilia Long, a Toronto advertising executive, warned of the dangers out on the sales floor: "Flattery will get you somewhere but over-familiarity will get you nowhere."[5]

A woman shopping for a stove was appraising a cooking appliance and trying to estimate future cooking performance. A man trying to sell her a stove had to step outside the prevailing masculine role, must suppress his inclination to explain how the stove was produced, and instead present himself as an authority on what the stove would produce, that is, on cooking. Manufacturers organized cooking schools for salesmen, although these attempts were usually reported as comical failures. In its "Use Value" campaign of 1950–51 the most Moffat's could persuade men to demonstrate was that water poured on top of the range would drain to the spill tray underneath. As one wag noted, the closest the appliance dealer, housewares manager, or salesman "ever came to basting a chicken was when he spilled champagne on one at the annual company dinner!"[6]

Reckoning that salesmen knew their culinary expertise would not bear scrutiny by an experienced homemaker, manufacturers tried to bolster male confidence by conjuring up a female customer who would not be a threat. They urged salesmen to imagine a dewy customer shopping for "the first range in her young life, or her first Moffat range." Canadian Westinghouse conjured up a new bride and in "An open letter to Westinghouse dealers" had her groom reassure salesmen, "There is nobody in this town who knows more about electrical appliances than you do—and, strictly between ourselves, there is nobody who knows *less* than my bride-to-be," and dismiss the threat of countervailing female authorities by asserting, "Like most newly weds my wife is going to be too proud or too shy to ask for advice from her friends or family—but if she feels that you, gentlemen, can guide her in the proper use of home appliances and help her avoid humiliating cooking failures, I think she will probably look upon you as her friend, and will probably buy from you again." Illustrations in sales guides portrayed difficult customers as older, ugly, and overbearing and good sales prospects as younger, attractive, and generously obliging. In these ways salesmen's fears of their most knowledgeable and discerning female customers were transformed into a negative commentary on the womanliness of the most skilled women shoppers.[7]

And in fact makers believed that they had more to gain by challenging women's traditional cooking knowledge than by making salesmen competent cooks in the estimation of women customers. In a competitive market each manufacturer wanted to redefine cooking as the satisfactory *operation* of its product, to emphasize recently purchased equipment rather than knowledge of ingredients or preparation skills as the keys to "baking and roasting success." The rhetoric in advertisements redefined cooking as a competitive, capital-intensive process. Manufacturers and retailers thus sought to shift credit for fine cookery from the expertise of the housewife to the engineering of the appliance.[8]

Sales pitches entailed a tactful but firm derogation of what women already knew about cooking compared with what they would need to know to get the most out of their new investment. Women cooks must be persuaded to subordinate themselves to their man-made equipment. The salesman's job was to "impress upon her [the woman customer] that the full benefit of all the wonderful things the range can do, all the work it can save, all the perfect cooking it can achieve are directly influenced by the way the range is handled. This is especially important with housewives who have long-standing cooking habits based on the use of old-fashioned or non-electric ranges. They have to be shown, and given confidence, in how much their Westinghouse Range will do for them—how little they need do themselves." Most makers included books of recipes from experts in their "test kitchens," said to be specifically adapted for their stove models, and then interleaved the recipes with instructions for operation of their equipment. Women journalists doubted that these tactics would succeed. Abbie Lane, woman's editor of the Halifax *Chronicle and Star,* told Ontario marketers that homemakers would not accept "the test kitchen and the home economist as the originator of recipes," that for "ordinary folks" recipes "have been in the family for years and that means something." Other women flatly denied that what they did in the kitchen was manage an investment.[9]

The elaborate accessorizing of early 1950s stoves, like the scientifically tested recipe, had more to do with selling than with cooking. The accessories at this stage were a way for manufacturers to differentiate one brand of white metal box from another. (Later in the decade they would begin to try to get consumers personally to identify with these distinctions.) More important for the salesman, the features provided content for the sales pitch; a stove with twice as many features had "twice as much for you to talk about—dozens of special sales features which mean added advantage to you [the salesman] in the days ahead." But the elaboration of features could undercut the promise that new equipment was easy to use. American studies from later in the decade suggested that retailers thought customers bought a product for its extra gadgets and style but that in fact customers were most interested in ease of use and in service and warranty. D. B. Cruickshank, an Ottawa businessman, worried about this contradiction in 1949: "Evidently the average householder's" choice is limited to what the average shop offers. . . . It appears to be sales managers' standards which are being forced on the consumer because the latter has no power of individual choice."[10]

The life-size cardboard cutouts of "attractive models" in strapless evening dresses that female shoppers found propped up against stoves in the stores and the couturier fashion shows that manufacturers staged amidst their appliance displays at the Toronto Canadian National Exhibition were planned to bring a positive association to stylishness in stoves. Thor's Canadian branch tried historical fashion shows as well, hoping to set women "reflecting how much better is their lot to be living in a day of simpler and saner fashions—and in a day when manufacturing enterprise removed so much of the household drudgery that plagued their grandmothers." Westinghouse combined the style and labor-saving messages in a before-and-after sequence in a rural setting. "Before" showed a harried farm wife feeding pancakes to a grinning foursome of hired men; "after" offered a neatly coifed woman playing bridge, smoking, and gossiping with three women friends. The wooden kitchen chairs are replaced by Breuer chrome, the wall display of the family patriarch and Lenten bulrushes by a northern landscape, the wood stove by a new Westinghouse electric. Let us count the ways this pitch would have failed: turning out grandfather, and the church, and the working men of the farm and replacing them with a northern wasteland, female leisure, and a baby crawling unremarked toward the stove.[11]

Women working at home for free were sensitive to criticisms that their days were unproductive and that household tasks required little skill.[12] They were not won over by claims about "how little they need do themselves" once they had a new range, partly, as a woman marketer noted, "because they resent being told that a product can do their job better and quicker than they can."[13] An appliance that "saved" the labor of a woman who did not work for wages cost, but did not save, money. Women whose home work was unmeasured and unpaid were not well positioned to press for purchases on the basis of labor-saving features even if they could see ways in which the new equipment would improve their work efficiency. As General Steel Wares discovered by "scientific market research" in 1954 as they planned to launch their new McClary ranges, "Most housewives would like to own a new automatic range but they feel the money could be better spent on other things of interest to the whole family."[14]

Many factors, then, made the appliance sales floor a bad place to shop for a stove. A cook could not try out the performance of the range there even in the limited way that a driver visiting an automobile dealership could test-drive a car. The men selling stoves were not experts in their use either in their own estimation or in that of their women customers. The manufacturers who provided the sales information about their ranges wrote copy to distinguish their product from that made by rival firms. They sold by looking into a mirror that reflected back other producers rather than by venturing through the looking glass to the distinctive world of the buyers beyond; they concentrated on the men who were their competitors rather than the women who were their customers.[15] Retailers sold the idea that new equipment was better than old, following the manufacturers' scripts, which emphasized features rather than functioning, attempting to sharpen the distinction between new and old by heightening and justifying sensitivity to superficial styling. Nowhere in this process were the questions the female shopper formulated as a cook satisfactorily addressed. Women protested that they got "the brush-off" when they went to research their prospective purchase (i.e., to shop), that marketers "slap a great big illustration and just a little reading matter" on their advertisements, "whereas when we really want to make a big purchase like this women will read advertisements packed with information."[16]

In response to the irrelevance of the sales floor to the real work of shopping for a stove, women buyers relied on word of mouth; they inquired of neighbors and kin how well the equipment they had purchased was meeting their needs, a route that offered information about the performance of the equipment in the home as long as acquaintances questioned were willing to be candid about their mistakes. Retailers, exasperated, claimed that all shoppers cared about was price. More likely, all the shopping for value and performance had been done through more reliable, nonretail routes. At the point of sale the only questions worth posing of a salesman about stoves concerned price.[17]

Designer Stoves

The processes by which stoves were designed were as far removed from the kitchen as was the appliance sales floor and equally revealing about the ideology of gender roles in the postwar years. The leading domestic appliance manufacturers had long histories as builders of industrial equipment, turbines for hydroelectric power generation, and giant boilers for steam plants. In this work the consumers were sovereign. Manufacturers of producer goods designed to the buyer's specifications. As Robert Campbell observed in *Canadian Business,* "The equipment would be built the way the customer wanted it and not the way the manufacturer thought it ought to be." Yet he observed wryly that large engineering firms setting out to make household equipment departed from this time-honored convention, assuming that their new client, the housewife, would know "that her humble role is to select from what is offered and not to advance opinions on what she'd like or why."[18]

Even though thousands of Ontario women had worked in heavy industry during the war, and some, the most well known being the aeronautical engineer Elsie Gregory Macgill, had made important design contributions to forward the war effort, in the urgent postwar national struggle to reconstruct the economy and create new industries through which to recoup Depression losses the masculinity of machine making was reconstructed unmodified as well. Honoring machine makers on their own terms became a species of patriotism; the unquestioning faith in the possibilities of the new technologies and materials developed for military purposes became a gesture of faith that peacetime prosperity could be recouped from the wreckage of war. Alan Jarvis, an influential figure in both British and Canadian industrial design after the war, used the example of the Mosquito bomber to make this case: "When we go shopping for those things with which we will fill our post-war homes, we ought, therefore, to remember the back-room boys who are trying to solve in the field of industrial design the same *kind* of problem which faced the designer of the Mosquito: how to make use of modern inventions, modern methods of production and newly discovered materials, so that we of the twentieth century shall have in our homes objects which are efficient, inexpensive *and* beautiful." In the area of consumer goods governmental industrial-design initiatives after the war emphasized the use of new materials, among these light metal alloys, and "the immediate possibilities of Canadian production" rather than the functional efficiency of equipment in the home. From the point of view of the Department of Finance, of course, this emphasis made perfect sense. The goal was to maximize sales at home and abroad of new materials, as well as goods made from those materials, gains which counted as growth in the gross national product. By contrast, efficiency changes inside the home were by definition outside the reckoning of the national income accounts. The comedic aspects of giving the production engineer free rein wherever he might roam did not go unremarked, but both state and industrial strategic priorities focused upon "modern" materials and production methods, as defined by men. In this sense, as Stuart Ewen has argued, form followed power, in this case the power to specify which gains forwarded the national interest.[19]

Functionalism, the assertion that form must follow function, was one of the rallying cries of postwar industrial design, but functionalism within the modernist's creed did not make the user paramount. Even in a publication titled *Design for Use* Donald W. Buchanan, the head of the National Industrial Design Council (later the National Industrial Design Committee, NIDC), argued that "the hall mark of any well designed, industrially produced object" was a "combination of functional efficiency and due proportion in form and structure." Materials and production processes had most direct influence on form and structure. Functional efficiency, which particularly in household equipment would have required attentiveness to the needs of women workers in domestic settings, consistently lost out in the designer's arbitrage among form, structure, and usefulness. Some even argued that a product was functional if it looked "its part," if in the contemporary system of signs it was possible "to recognize actually what it is." In the criteria for industrial design awards formulated a year after the *Design for Use* statement, Buchanan and his colleagues did not demand functional efficiency; they required only that objects be "suitable" rather than optimal for their functions. Suitability implied "further that the object be both comfortable and easy to handle," an amplification that evermore drew attention to form rather performance, to the object itself rather than the task to be done.[20]

Retailers were perplexed that articles that "had widely established consumer acceptance and appeared to be excellent value" were ranked low by the council's design experts "on account of lack of originality of form." American industrial designers who valued sales appeal tended to leaven the strict formalism that the British theorists whom Buchanan and his colleagues in the national design bureaucracy so admired. But in this leavening the users' needs did not rise to the top. Henry Dreyfuss, who designed appliances for General Electric in the United States, cheerfully argued before the Canadian Manufacturers Association in Toronto in May 1952 that it was a good thing that industrial design had "entered the home through the back door," into the kitchen, where "wear and tear were faster," and the housewife, "a gadget-conscious mammal," could be persuaded to "have the drab parts of her house brightened-up with handsome bits of machinery." Dreyfuss found women users too mercurial, or disingenuous, or hedged in by their relationships to the men in their lives to provide reliable information about what they really wanted. An infrequent user of household appliances himself, he dismissed as retrograde instincts the labor patterns women had developed to manage their work in the kitchen. Monte Kwinter, in the 1950s the managing editor of a Toronto-based product-design journal and later the minister of consumer and corporate affairs in Ontario, stated this view even more strongly, arguing that "it would be advantageous to General Electric and to industry as a whole to educate people to understand what they should want . . . in other words, to give them what they need rather than what they want."[21]

The Canadian Association of Consumers (CAC), whose membership at the time was restricted to women and which worked closely with the National Industrial Design Council during the early 1950s, took a contrary view. Under the leadership of women with close links to the governing federal Liberal Party, the association pursued two goals, to represent the interest of women consumers through briefs to government and industry and to urge Canadian women to be discriminating shoppers, making the market system work for them by refusing to buy goods that did not meet their needs.[22]

In 1946 its early leadership had developed a bolder plan, hoping to build upon their experience with the all-women Consumers Branch of the Wartime Prices and Trade Board. Drawing together suggestions from a national survey of the Local Councils of Women and Consumer Councils, Mrs. W. R. Lang, an Ottawa resident who represented the Women's Missionary Society of the United Church, reported that women across the country wanted a consumer research bureau that would "help consumers to crystallize out their needs and desires," which they acknowledged were, because they were without engineering knowledge, "often vague and inarticulate." This agency, perhaps within the federal Department of Trade and Commerce, would bring "great benefits not only to the consumer but also to the manufacturers" by conducting basic research into "the necessary activities of housewives" and how various kinds of equipment would "affect the time and energy of those using them" and then formulating specifications that would "express the needs and wishes of the consumer before things are manufactured." In the same way that the National Research Council had the job of reinterpreting basic science for Canadian industrialists, this body would bridge the gap between women consumers' needs and their ability to formulate authoritative engineering specifications, work the British Institute of Planning and Design currently was engaged in and that the Swedish Home Research had undertaken, on the initiative of leading Social Democratic women, since 1942.[23] This plan was intended to better women's place in the world of goods by scaling through state intervention three scarcely acknowledged but almost insurmountable barriers that the contemporary system of gender relations put in their way: it would have created productivity data for household work, subverted the male monopoly over engineering solutions to technical problems women experienced within the home, and presented manufacturers with design alternatives that genuinely were functional and by this fact alone need not have been unsalable.

The plan never was realized. The consumers' association that emerged from the contending models proposed by communist, social democratic, and liberal women in the late 1940s relied upon a small and uncertain federal grant to fund its activities. Rather than following the National Research Council precedent, the women of the CAC worked as women worked within the home, using unpaid labor, saving and sharing information as it came to hand, hoping through astute petitioning and wise purchasing to minimize the irrationality of the choices the market inattentively put before them.

Work with the NIDC put the CAC's optimist liberal principles to a stern test. The CAC's concern from the start was with poor performance in household goods. They began by recruiting suggestions from members, which they expected the NIDC to forward to "departments of household science for investigation in report," a process they hoped would allow women, "instead of stewing over an inconvenience and feeling helpless to improve upon it," to complain constructively to "a source which will take immediate action" and to create "findings which it will be worthwhile giving to manufacturers."[24]

For its part, the NIDC hoped to convert consumers to modern design principles. They began with a pamphlet, *It Pays to Buy Articles of Good Design,* mailed out to all CAC members, which by annotated picture comparisons praised goods that were simple rather than ornamented, revealing rather than hiding their functions, "natural" in the materials from which they were made rather than imitative of materials from which they were not.[25]

A member of the CAC agreed to served on the jury for the annual design awards, established in 1953, and here the conflicts between consumers' and designers' priorities quickly crystallized. Awards regularly were given to goods the woman juror and her committee thought showed evident faults and withheld from wares they favored. The award insignia created the impression that products had been performance tested when they had not. Aware that their participation in the judging validated this misimpression, in 1954 and 1955 Kathleen Harrison, chair of the CAC Design Committee, tried to persuade the council to separate household equipment from decorator goods and withhold awards until the equipment was tested by an established research group. In the meantime members of the Ottawa branch began to test goods in their homes, although their work had to be limited to small electrical units and housewares, and the volume of the task soon overwhelmed the small group of available volunteers. The NIDC declined to make performance testing part of their own adjudication process or to require certification of performance testing from manufacturers. Their only response was an extraneous change in the award regulations to require "evidence of suitability for the Canadian market," that is, of sales rather than usefulness. Kathleen Harrison thought she had arranged for a public airing of the dispute through a "provocative" article in *Canadian Homes and Gardens* entitled "How Good Is Design Award Merchandise?" but the editorship of the journal changed and the article never appeared. Their experience with the award jury taught the female CAC that designers and industrialists often wanted their validation rather than their contribution. At the NIDC a confidential memo linked the breach to the fact that "the male of the species is not considered a consumer in the CAC!"[26]

A Women's Stove

The women of Ontario had strong views about stoves. When the NIDC and the CAC in 1954 jointly requested suggestions for the improvement of household goods, two-thirds of the responses concerned kitchen equipment.[27]

On one hand, women, disenchanted with the existing merchandising system, wanted better ways to get the product knowledge they needed in order to choose among the stoves manufacturers offered for sale. This meant, first, that they wanted fewer model changes. When appliance manufacturers, following the example of automobile makers, changed their designs every year, they removed from the market the very models, now proven by two or three years' performance in the kitchens of family and friends, that women wanted most to buy. "Why," one woman wrote, do manufacturers not recognize that they would be better off adopting "a wait-and-see attitude rather than permit themselves to be dazzled by the immediate selling power of novelty into scrapping models which may be inherently better, as the consumer will, in time, come to recognize?" Second, worn down by "the sheer physical effort of shopping," and aware that smaller manufacturers did not have the advertising budgets or the clout with retailers to secure due attention for their products, women suggested that makers "establish cooperative demonstration rooms" to ensure "that *everybody's* product could be seen and inspected by prospective purchasers." Showing a sharp sense of the differences between oligopoly and perfect competition, one woman argued that this change actually would make the market process, which was supposed to allow the best goods to rise to the top of consumer acceptance, work better by ensuring that "the smaller manufacturer gets every assistance in his struggle to compete."[28]

On the other hand, women wanted the form of the stoves manufacturers made changed in a fundamental way. They wanted the ovens raised above, in their own terms "waist level," in designers' terms "counter level." "Why, oh why," wrote Mrs. R. F. Legget of Ottawa, "do manufacturers persist in placing on the market stoves that are beautiful to look at, with the chromium glistening enamel, but back-breaking to use?"[29] This was a longstanding female complaint in Ontario. In 1946 the chrome-plated modern tabletop range included in the Art Gallery of Toronto "Design in the Household" exhibition drew much criticism. Visitors' questionnaires revealed that consumers wanted "the oven [to] be restored to its old position above the stove. Housewives find that stooping over to attend to baking is an unwelcome form of exercise."[30]

Historically, in the most common Canadian ranges, burning solid fuel, ovens had been set beside rather than below the burners. Early gas and electric stoves followed this form. Donald Buchanan, skeptical when he could find only one Canadian electrical range, a McClary, with the oven at what the housewives claimed was the right height, referred the matter to an authority, Dr. J. B. Brodie, head of household science at the University of Toronto. She made common cause with the housewives, arguing that plans "to 'streamline' everything and have a working space around the kitchen at one level . . . are evolved by those who do not work in a kitchen and we know that they are not efficient." Her words were well chosen. Buchanan, a proponent of British Good Design principles, regarded streamlining as a heresy hatched by American salesmen. He circulated Brodie's report to makers in the hope that "some enterprising manufacturer will devise a better and more functional type of range, that will appeal to the consumer on the grounds of both *appearance* and *ease of handling*" and through a colleague assigned a group of University of Toronto architecture students to look at the problems of range design "afresh, without preconceived prejudices as to what is 'modern.'" But both the students' mock-up and an ungainly one-legged English-style cooker, designed to preserve the plane geometry of the countertop, met with lukewarm responses from members of the public who participated in design quizzes at Toronto and Ottawa exhibitions in the late forties.[31]

When 75 percent of CAC members surveyed in 1951 still favored high-oven ranges, Mrs. R. G. Morningstar, of Toronto, the CAC representative from the Canadian Dietetic Association, was dispatched to interview manufacturers. The results were not encouraging. Although the high-oven mock-ups were no wider than the stoves currently in production, manufacturers declared them too massive. They cited the failures of high-oven models in American market tests. Mr. Alexander McKenzie, at McClary's, added that "he felt that older women were not familiar with modern controls" and that when using a high oven they would "keep peeking into the oven to see how the product is baking."[32]

One Ontario firm, Findlay Stove Manufacturers of Carleton Place, did agree to take up the project. They exhibited a sample high-oven stove at the Toronto exhibition in 1952 and over the next three years conducted surveys of their own and refined their design, working in close concert with the CAC Design Committee. The Findlay Hy-Oven was launched in 1955 and that year duly won a National Industrial Design Award. The CAC publicity chair in Toronto wrote letters to 279 newspaper and magazine editors and women's radio commentators under the banner, "From Modernity to Convenience" and garnered 1,666 lines of newspaper publicity. Findlay's ad copy noted that the Hy-Oven ended stooping and bending and took up only 41.5 inches of floor space and with floor plans demonstrated how the new model would fit conveniently into existing kitchens. But the Hy-Oven did not sell, either in Canadian or export markets.[33]

Findlay's itself may have been partly responsible for the failure of the Findlay Hy-Oven. A small if venerable eastern Ontario firm, Findlay's may have lacked the market influence and the advertising budget required to successfully launch any radically different product in a sector increasingly dominated by large American firms. But there was also a problem with the high-oven design; in some ways its usefulness compromised the ability of the form to satisfy other consumer needs. Mrs. G. J. Wherrett, CAC Design Committee chair, in 1952 reported rumors that women who saw the Hy-Oven at the Toronto Canadian National Exhibition "liked the convenience of the high oven, but wanted the appearance of the low oven." In her 1954 NIDC-CAC contest entry Mrs. John M. Dexter, of Burritt's Rapids, Ontario, said that she still held the view that women preferred "the old high-oven stove to the most modern, streamlined one," but attending to "manufacturers" complaints that the high-oven stove does not lend itself to good design in a modern kitchens, she offered a compromise sketch of a range with a retractable oven that could be raised while in use and stowed below counter level at other times.

Henry Dreyfuss argued that American high-oven revivals had failed in the market because "the table-top stove flush with the other cabinets in the kitchen had become such a style factor that the ladies refused to be budged away from it."[34] The CAC pitch for the Hy-Oven—"From Modernity to Convenience"—posited too stark a transition. Modernity was not irrelevant to women shopping for convenience. Women were shopping for stoves *and* for the characteristics the presence of a particular stove would bring to their kitchen. They were unlikely to see, as they imagined a stove transported from the sales room to their kitchens, exactly the same transformation that the designer, the manufacturer, and the advertiser, each in turn, had envisioned. It is not that any dysfunctional contraption offered up for sale would have found acceptance with women buyers simply because it was perceived as creating the right cultural effect. But for the woman shopper, as for all the men who had created the array of products from which she would choose, a stove was more than just a stove.[35]

The tabletop range, its work surface level with the surrounding counters, was a central feature in the portrait of the kitchen as laboratory, an image that Margaret Hobbs and Ruth Pierson have shown was expounded widely in Canada in the 1930s.[36] Its enamel sparkling cleanliness, its clock and timer and array of knobs promising orderliness and scientific control, the range was the necessity among "the electrical tools of her trade," which radio announcers for Canadian Westinghouse in 1951 proclaimed were "rapidly changing the homemaker's kitchen into a convenient, attractive workshop." In the postwar years the laboratory kitchen signaled the worth of domestic work to women leaving the paid labor force and the feasibility of combining waged and nonwaged work for those who were not.[37] The messages were not unmixed, for the unyielding regularity of the laboratory aesthetic brought home the imagery of the factory or the office and the sense of being controlled rather than being in control that some women returning to the home had hoped to escape.[38]

Later in the decade the once-admired laboratory look in kitchens was derogated as the "utility" look, a characterization that in Ontario brought to mind the austere minimalism of wartime goods.[39] Wealthier householders then installed conveniently high wall ovens in their custom-designed kitchens and traded in the tabletop ranges that by the late fifties seemed "to reflect a greater pre-occupation with the romance of science than with the comforts of home."[40] But at the time when the men at Findlay's and the women of the CAC were trying to launch their Hy-Oven, they were struggling against not only a machine makers' practice and a national economic strategy that did not take domestic labor into account but also a generation of consumers seeking the signs of science to validate their unpaid work within the home.

This is a story that can only be told of Ontario in the early 1950s. Then the wartime experience of women in the Consumers Branch and the relatively unformed practices of mass consumption in domestic durable goods gave some women confidence to plan ways to shape the design and merchandising of the equipment they would use. The persistence of small Canadian foundries, accustomed to manufacturing in short runs for specialized markets, meant that there was at any rate one local firm ready and able to undertake the experiment. By the end of the decade the small Ontario manufacturers had lost their independence. The CAC had abandoned its attempt to specify the form of household equipment before it was manufactured and had begun instead to prepare buyer's guides to help shoppers sort their way through the goods that manufacturers and retailers offered up for sale. The ranks of the appliance salesmen dwindled, their usefulness as intermediaries across the gender gap between makers and consumers disproved in the hectic postwar boom. By the late 1950s Ontario appliance retailing was dominated by suburban discounters with small staffs to write up bills of sale for hopeful if weary women who disembarked from station wagons shopping for a good stove.

Notes

This chapter was written while I was a Bunting Fellow at Radcliffe College. I am grateful to Jehan Kuhn, Eve Blau, and Alice Friedman for comments on the early drafts and to audiences at the Hagley Museum and Library (Wilmington DE), Northern Lights College (Fort Nelson BC), and Malaspina College (Duncan BC) and to colleagues in the Toronto Labor Studies Group for critiques of subsequent versions. It was previously published in Joy Parr, ed., *A Diversity of Women: Ontario, 1945–1980* (Toronto: Univ. of Toronto Press, 1995), 75–97, and is reprinted here with minor changes by permission.

1. "Cultivate the Ladies," *Hardware in Canada,* Aug. 1946, 26, 28; Frank Wright, "The Woman Nation's Purchasing Agent," *Canadian Business,* Aug. 1946, 132; Archives of Ontario, T. Eaton Company, S69, Merchandise Office, vol. 12, "Business Conditions and Forecasts, 1940–52"; B. E. Mercer to W. Park, 9 Feb. 1950, summarizing talk at a meeting of the Retail Federation by Edythe Fern Melrose, and article in *Hardware Metal and Electrical Dealer,* 4 Feb. 1950. The study from which the data Mercer and Park used were drawn was published by Paul Converse and Merle Crawford as "Family Buying: Who Does It? Who Influences It?" in *Current Economic Comment,* 11 Nov. 1949, 38–50. Mirra Komarovsky compared a number of postwar studies showing similar outcomes in "Class Differences in Family Decisionmaking on Expenditures," in *Household Decisionmaking,* ed. Nelson N. Foote (New York: New York Univ. Press, 1961): 255–65. On the male shadow see Wright, "The Woman Nation's Purchasing Agent"; on the impatient husband see "Mrs. Consumer Speaks Out on Design," *Industrial Canada,* Jan. 1955, 78. The literature on women's level of autonomy in various types of decision making is well summarized in Rosemary Scott, *The Female Consumer* (New York: John Wiley, 1976): 119–25.

2. For positive views of women in appliance sales see McMaster University Special Collections (McMU), Moffat's *Sales Chef,* Jan.–Feb. 1952, commentaries by C. A. Winder, general sales manager, and the column "The Woman's Angle," by Elaine Collet. The negative view is from a male sales consultant, "More Selling, Less Crying," *Marketing,* 18 Mar. 1955, 24. *Marketing* was the leading Canadian trade weekly in its field. For contending views on why commission appliance sales is a men's job in the United States see "Women's History Goes on Trial: EEOC v. Sears, Roebuck and Company," *Signs* 11 (summer 1986): 751–79; and Ruth Milkman, "Women's History and the Sears Case," *Feminist Studies* 12 (summer 1986): 375–400.

3. See, e.g., "Eight Sound Rules for Selling," *Westinghouse Salesman,* May 1948, 2, in McMU, Westinghouse Collection, box 7, file 27; and on Moffat's, "Virile

Salesmanship and Advertising Built World-wide Canadian Business," *Marketing,* 2 Oct. 1948, 56.

4. Charles Pearce, "Watch Your Language!" and "Why True-temp Is Better," *Westinghouse Salesman,* July 1947; G. I. Harrison, "I Lost the Sale Because . . . ," ibid., June 1948; "The Woman's Angle," *Sales Chef,* Mar. 1951; "Would-be Purchasers Still Getting Brush-off, the Customer Declares," *Marketing,* 13 Apr. 1946, 10.

5. "The Sales Clinic," *Canadian Business,* Jan. 1948, 92; Harrison, "I Lost the Sale Because . . . ," 9; "Life with Buy-ology," *Westinghouse Salesman,* July 1951, 8–9; "Selling via Demonstrations," *Marketing,* 30 July 1949, 8; "Manufacturer's Retailers' Guide Promotes Many Competing Products," ibid., 16 Sept. 1950, 18; "Canadian Women Want Product Facts," ibid., 12 Dec. 1951, 18.

6. Robert M. Campbell, "You Can't Wash Dishes with Aesthetics!" *Canadian Business,* June 1951, 41.

7. *Sales Chef,* Dec. 1950, 5; "Women's Views of Moffat News," ibid., Nov. 1948; *Westinghouse Salesman,* Apr. 1951, 11; *Sales Chef,* July 1950, 7, and Mar.–Apr. 1950, 6; Campbell, "You Can't Wash Dishes with Aesthetics!" 41.

8. On selling an appliance as a redefinition of a task see Carolyn Shaw Bell, *Consumer Choice in the American Economy* (New York: Random House, 1967), 224; *Sales Chef,* Nov. 1948, Feb.–Mar. 1951, 17, July–Aug. 1947. On this pattern in interwar Britain see Suzette Worden, "Powerful Women: Electricity in the Home, 1919–40," in *A View from the Interior: Feminism, Women, and Design,* ed. Judy Attfield and Pat Kirkham (London: Women's Press, 1989). Keith Walden describes a similar campaign by manufacturers to turn attention from local skills toward goods sold in national markets in "Speaking Modern: Language, Culture, and Hegemony in Grocery Window Displays, 1887–1920," *Canadian Historical Review* 70 (Sept. 1989): 296, 308–9.

9. "Help Yourself to Future Range Sales," *Westinghouse Salesman,* Sept. 1950 (quotation); "Practical Ideas for Advertisers in Cultivating Women Customers," *Marketing,* 11 Dec. 1948, 14; "Would-be Purchasers Still Getting Brush-off, This Customer Declares," ibid., 13 Apr. 1946, 10.

10. "Service Page," *Sales Chef,* Sept. and Nov. 1948; *Marketing,* 30 Nov. 1948, 18; Peter J. McClure and John K. Ryans, "Differences between Retailers' and Consumers' Perceptions," *Journal of Marketing Research* 5 (Feb. 1968): 37; D. B. Cruickshank, "Industrial Design, What We Are Doing About It," speech, c. 7 June 1949, in National Gallery of Canada, Ottawa (hereafter NG), vol. 7.4, "Design in Industry," file 2; Scott, *Female Consumer,* 67–68. On appliance gadgets and sales in other settings see Judy Wajcman, *Feminism Confronts Technology* (University Park: Pennsylvania State Univ. Press, 1991), 103–4; and T. A. B. Corley, *Domestic Electrical Appliances* (London: Jonathon Cape, 1966), 136.

11. *Sales Chef,* Mar.–Apr. 1950, Nov. 1952; *Marketing,* 15 Aug. 1953, 3; "The Farmer's Wife," *Westinghouse Salesman,* Feb. 1948, 9.

12. See, e.g., Norton Calder, "Women Are Lousy Housekeepers," *Liberty*, Nov. 1954, 15; and "Wasted Effort," *Saturday Night*, 30 Oct. 1954, 3.

13. "Women Best Ad Readers, Shrewder Buyers," *Marketing*, 29 Apr. 1955, 8.

14. "50 Percent of Ranges 10 Years Old Basis for McClary Campaign," ibid., 8 Oct. 1954, 6, 8. Dianne Dodd notes that in the interwar period in Canada advertisers responded to this concern by emphasizing the benefits of labor-saving domestic appliances to the whole family (Dodd, "Women in Advertising: The Role of Canadian Women in the Promotion of Domestic Electrical Technology in the Interwar Period," in *Despite the Odds*, ed. Marianne Ainley [Montreal: Vehicule, 1990], 144–45). Corley suggested similarly in the British case that women may have feared being thought lazy if they had too many appliances, or they may have preferred to agree that money be spent on items more immediately pleasing to their husbands (Corley, *Domestic Electrical Appliances*, 133; see also Vance Packard, *Hidden Persuaders* [New York: David McKay, 1957], 62, and Maxine L. Margolis, *Mothers and Such* [Berkeley: Univ. of California Press, 1984], 167).

15. On this point see Harrison White, "Where Do Markets Come From?" *American Journal of Sociology* 87 (Nov. 1981): 543–44; and Susan Strasser, *Satisfaction Guaranteed: The Making of the American Mass Market* (New York: Pantheon, 1989), 289.

16. See the following articles in *Marketing:* "Would-Be Purchasers Still Getting Brush-off"; "Canadian Women Want Product Facts"; and "Women's Purchasing Viewpoints Explained . . . ," 19 Jan. 1952, 4.

17. "Charting the Course for Selling," *Marketing*, 30 Nov. 1946, 2; "Selling via Demonstrations," 12; *Westinghouse Salesman*, July 1951, 9; "Mrs. Consumer Speaks Out on Design." William H. Whyte invokes this dilemma well in "The Web of Word of Mouth," *Fortune* 50 (Nov. 1954), reprinted in *Consumer Behavior*, vol. 2 (New York: New York Univ. Press, 1955), 118–20.

18. Campbell, "You Can't Wash Dishes With Aesthetics!" 40–41. For similar patterns in Britain and Australia see Corley, *Domestic Electrical Appliances*, 52; and Wajcman, *Feminism Confronts Technology*, 100–103.

19. "What's All This Fuss about Modern Design?" *Target*, 9 Oct. 1945, 10, in Thomas Fisher Rare Book Room, Robarts Library, University of Toronto, Alan Jarvis Collection, ms. 171, box 22; Statement by the Affiliation of Industrial Designers of Canada, 3 July 1947, NG, vol. 7.4, "Design in Industry," file 1; speech by C. D. Howe, minister of reconstruction, opening Design in Industry Exhibition, 1946, ibid., vol. 5.5, file D; John B. Collins, "Design for Use, Design for the Millions: Proposals and Options for the National Industrial Design Council, 1848–1960," (master's thesis, Carleton University, 1986), ch. 2; *Westinghouse News*, 2 June 1948; Stuart Ewen, *All Consuming Images* (New York: Basic Books, 1988), ch. 9.

20. NIDC, *Design for Use* (Ottawa: King's Printer, 1947), preface by Donald

Buchanan, 5; on acting the part see Dora de Pedery to H. O. McCurry, 23 Sept. 1948, NG, vol. 5.5, file D, "Design Centre Exhibition 1948"; *It Pays to Buy Articles of Good Design* (Ottawa: National Gallery and CAC, n.d.); *Industrial Canada*, Jan. 1948, 141; Donald Buchanan, "Completing the Pattern of Modern Living," *Canadian Art* 6 (spring 1949): 112; *Foreign Trade*, 25 Nov. 1950, 898.

21. B. W. Smith of Eaton's Merchandising Office, 25 Feb. 1953, Archives of Ontario, T. Eaton Company, S69, box 19, "Design Index and Awards, 1946–60"; Henry Dreyfuss, "The Silent Salesman of Industry," *Industrial Canada*, July 1952, 65; idem, *Designing for People* (New York: Simon & Schuster, 1955), 65–66, 202; Monte Kwinter, quotation from commentary on paper by C. H. Linder at NIDC, "Design as a Function of Management," 18 Oct. 1956, National Archives of Canada (NAC), RG 20 A4 1434, file 2.

22. CAC, *Your Questions and Answers* (Ottawa, 1947), question 9; address by Mrs. W. R. Walton, President, CAC, to the Toronto Business and Professional Women's Club, NIDC press release, 22 Dec. 1952, NG, vol. 7.4, "Design in Industry," file 5; Mrs. Rene Vautelet, presidential address to CAC's 1954 annual meeting, NAC, Manuscript Group (MG) 28 I 200, vol. 1.

23. Mrs. W. R. Lang, "Report re questionnaire, May 3, 1946," Metropolitan Toronto Reference Library, Baldwin Room, Harriet Parsons Papers.

24. Minutes of the National Industrial Design Committee, 23 and 24 Apr. 1951, NAC, RG 20 997, vol. 2; extracts from the annual meeting of the CAC, 29 Sept. 1954, NAC, RG 20, vol. 1429, National Design Branch and Consumers Association of Canada, file 1000 240/C27.

25. NIDC and CAC, *It Pays to Buy Articles of Good Design* (Ottawa, 1951); Collins, "Design for Use, Design for the Million," notes the didactic tone of this work.

26. CAC, Design Committee, midyear report, Apr. 1955, and report to annual meeting, 5, 6 Oct. 1955, NAC, MG 28 I 200, vol. 1; Mrs. K. E. Harrison, chair of the Design Committee, to Executive of CAC, 26 Feb. 1955, NG, vol. 7.4, "Design in Industry," file 7; Mrs. C. Breindahl, CAC, Ottawa, to Donald Buchanan, requesting performance testing, 15 Apr. 1955, and Buchanan to Breindahl, 20 Apr. 1955, describing the change in award specifications, and Harrison to Buchanan, 23 Apr. 1955, stating that volunteers are overburdened and do not feel they constitute a valid testing unit, all in NAC, RG 20, vol. 1429, file 1000 240/C27. This file also includes the home testing questionnaire formulated by the Ottawa branch and their April 1956 report on their work in 1955. An unsigned confidential memo for the consumer relations council of the NIDC written around October 1955 summarizes the history and work of the CAC from the perspective of the NIDC.

27. There were 896 responses in all, 366 from Ontario (NAC, RG 20, vol. 1429, file 1000 240/C27, "Review of the 1954 NIDC-CAC contests for suggestions for the improvement of household goods").

28. "Mrs. Consumer Speaks Out on Design," *Industrial Canada,* Jan. 1955, 76, 78, 80.

29. "What the Women Want," *Canadian Homes and Gardens,* Jan. 1955, 36; "Kitchen Opinion Polls Call Grandma's Oven Back Again," *Financial Post,* 2 Apr. 1955.

30. Barbara Swann, "The Design in the Household Exhibition." *Industrial Canada,* Mar. 1946, 77–78. There are excellent curatorial records describing this exhibition in the Art Gallery of Ontario archives.

31. Hillary Russell, "'Canadian Ways': An Introduction to Comparative Studies of Housework, Stoves and Diet in Great Britain and Canada," *Material History Bulletin* 19 (spring 1984): 1–12, a fine article containing clear illustrations of early Canadian high oven ranges; Buchanan to George Englesmith, 16 Sept., 14 Nov., and 21 Nov. 1947, NAC, RG 20 A4, vol. 1433, National Design Branch, file entitled "George Englesmith, architect"; Donald Buchanan, "Take Another Look at Your Kitchen Range," *Canadian Art* 5 (spring/summer 1948): 182–83; J. B. Craig, F. Dawes, and J. C. Rankin, "A Cooperative Problem in Industrial Design," *Royal Architectural Institute of Canada Journal* 25 (May 1948): 154–55. For Buchanan on streamlining see "Design in Industry: The Canadian Picture," ibid. 24 (July 1947): 234–35; and "Good Design and 'Styling'—The Choice before Us," *Canadian Art* 9 (Oct. 1951): 32–35. "These are the Ones the Experts Picked," ibid. 6 (Dec. 1948): 59–60, shows the University of Toronto mock-up and a British Maxwell Fry design. Cruickshank, "Industrial Design, What We Are Doing About It."

32. June Callwood, "Housewives' Crusade," *Macleans,* 1 Oct. 1952, 60; CAC, report of the Consumer Relations Committee, 20–21 Sept. 1951, NAC, MG 28 I 200, vol. 1, file 10; Mrs. R. G. Morningstar, "Survey of Time and Motion Studies for Household Equipment," report no. 23, 1952, ibid.; minutes of the NIDC meeting, 15–16 Apr. 1952, NG, vol. 7.4-D, "Design in Industry," file 4.

33. "Review of the 1954 NIDC-CAC Contest," 4, NAC, RG 20, vol. 1429, file 1000 240/C27; "From Modernity to Convenience," *CAC Bulletin,* Jan. 1955, in University of Guelph Archives, M. S. McCready Collection, A0021; "National Industrial Design Report," presented to CAC 1952 annual meeting by Mrs. G. J. Wherrett, and Design Committee report by Mrs. W. F. Harrison, Apr. 1955, 3–4 June 1955, and 5–6 Oct. 1955, NAC, MG 28 I 200, vol. 1; *Furniture and Furnishings,* Apr. 1955, 83. On early export performance see D. G. W. Douglas to J. P. C. Gauthier, Department of Trade and Commerce report, 15 July 1955, 5, NAC, RG 20, vol. 998, file 13-2.

34. Report by Mrs. G. J. Wherrett, Design Committee, to 1952 CAC annual meeting, NAC, MG 28 I 200, vol. 1; "What the Women Want," 36; Dreyfuss, *Designing for People,* 69.

35. On the characteristics of goods see Jean-Christophe Agnew, "The Consuming Vision of Henry James," in *The Culture of Consumption: Critical Essays in*

American History, 1880–1980, ed. Richard W. Fox and T. J. Jackson Lears (New York: Pantheon, 1983), 67–74. On the assertion of Jean Baudrillard that the real effect of consumption has been to herald "the passage from use value to sign value" see Alan Tomlinson, introduction to *Consumption, Identity and Style: Marketing, Meanings and the Packaging of Pleasure,* ed. Tomlinson (New York: Routledge, 1990), 18–21. On consumer reinterpretation of designs see Cheryl Buckley, "Made in Patriarchy: Toward a Feminist Analysis of Women and Design," *Design Issues* 3 (fall 1986): 11; and Phillipa Goodall, "Design and Gender," *Block* 9 (1983): 50–61. On reciprocity between the values goods bring to users and users bring to goods see Penny Sparke, *Electrical Appliances* (London: Unwin Hyman, 1987), 6; and Strasser, *Satisfaction Guaranteed,* 15.

36. Margaret Hobbs and Ruth Roach Pierson, "'A Kitchen That Wastes No Steps . . .': Gender, Class, and the Home Improvement Plan, 1936–40," *Histoire sociale—Social History* 21 (May 1988): 9–37.

37. "Canadian Westinghouse Presents," broadcast script for 11 Feb. 1951, in McMU, Westinghouse Collection, box 8, file 21; J. K. Edmonds, "The Mechanized Household," *Marketing,* 5 Nov. 1954, 17.

38. On the ambiguous messages carried by factory and laboratory imagery see Stuart Ewen, "Marketing Dreams: The Political Elements of Style," in Tomlinson, *Consumption, Identity, and Style,* 48–49; and idem, *All Consuming Images,* 215–16.

39. "Kitchens," *Canadian Homes and Gardens,* Oct. 1958, 30.

40. "Long Range Planning for Tomorrow's Kitchens," *Design* 104 (Aug. 1957): 50.

Eight

From Town Center to Shopping Center

The Reconfiguration of Community Marketplaces in Postwar America

Lizabeth Cohen

When the editors of *Time* magazine set out to tell readers in an early January 1965 cover story why the American economy had flourished during the previous year, they explained it in terms that had become the conventional wisdom of postwar America. The most prosperous twelve months ever, capping the country's fourth straight year of economic expansion, were attributable to the American consumer, "who continued spending as if there were no tomorrow." According to *Time*'s economics lesson, consumers, business, and government "created a nonvicious circle: spending created more production, production created wealth, wealth created more spending." In this simplified Keynesian model of economic growth "the consumer is the key to our economy." As R. H. Macy's board chair Jack Straus explained to *Time*'s readers, "When the country has a recession, it suffers not so much from problems of production as from problems of consumption." And in prosperous times like today "our economy keeps growing because our ability to consume is endless. The consumer goes on spending regardless of how many possessions he has. The luxuries of today are the necessities of tomorrow." A demand economy built on mass consumption had brought the United States out of the doldrums of the Great Depression and World War II, and its strength in the postwar period continued to impress those like retail magnate Straus, whose own financial future depended on it.[1]

Although Straus and his peers invested great energy and resources in developing new strategies for doing business in this mass-consumption economy, historians have paid far less attention to the restructuring of American commercial life in the postwar period than to the transformation of residential experience. An impressive literature documents the way the expansion of a mass consumer society encouraged a larger and broader spectrum of Americans to move into suburban communities after the war.[2] Between 1947 and 1953 alone the suburban population increased by 43 percent, in contrast to a general population increase of only 11 percent.[3] At an astonishing pace the futuristic highways and mass-built, appliance-equipped, single-family homes that had been previewed at the New York World's Fair in 1939–40 seemed to become a reality. Thanks to a shortage in urban housing, government subsidies for highway building and home construction or purchase, and pent-up consumer demand and savings, a new residential landscape began to take shape in metropolitan areas, with large numbers of people commuting into cities for work and then back to homes in the suburbs. (Increasingly as the postwar era progressed, suburbanites worked, not just lived, outside cities.)

Less explored by historians and slower to develop historically was the restructuring of the consumer marketplace that accompanied the suburbanization of residential life. New suburbanites who had grown up in urban neighborhoods, walking to corner stores and taking public transportation to shop downtown, were now contending with changed conditions. Only in the most ambitious suburban tracts built after the war did developers incorporate retail stores into their plans. In those cases developers tended to place the shopping district at the core of the residential community, much as it had been in the prewar planned community of Radburn, New Jersey, and in the earliest shopping centers, such as Kansas City's Country Club Plaza of the 1920s. These precedents, and their descendants in early postwar developments in Park Forest, Illinois, Levittown, New York, and Bergenfield, New Jersey, replicated the structure of the old-style urban community, where shopping was part of the public space at the settlement's core and residences spread outward from there.[4] But most new suburban home developers made no effort to provide for residents' commercial needs. Rather, suburbanites were expected to fend for themselves by driving to the existing market towns, which often offered the only commerce for miles, or by returning to the city to shop. Faced with slim retail offerings nearby, many new suburbanites of the 1940s and 1950s continued to depend on the city for major purchases, making do with the small, locally owned commercial outlets in neighboring towns only for minor needs.

It would not be until the mid-1950s that a new market structure appropriate to this suburbanized, mass-consumption society prevailed. Important precedents existed in the branch department stores and prototypical shopping centers constructed between the 1920s and 1940s in outlying city neighborhoods and in older suburban communities, which began the process of decentralizing retail dollars away from downtown. But now the scale was much larger. Even more significant, the absence or inadequacy of town centers at a time of enormous suburban population growth offered commercial developers a unique opportunity to reimagine community life with their private projects at its heart.[5]

By the early 1950s large merchandisers were aggressively reaching out to the new suburbanites, whose buying power was even greater than their numbers.[6] The 30 million people that *Fortune* magazine counted as suburban residents in 1953 represented 19 percent of the U.S. population but 29 percent of its income. They had higher median incomes and homeownership rates, as well as more children fourteen and under, than the rest of the metropolitan population, all indicators of high consumption.

Merchandisers also realized that postwar suburbanites were finally living the motorized existence that had been predicted for American society since the 1920s. As consumers became dependent on, virtually inseparable from, their cars, traffic congestion and parking problems discouraged commercial expansion in central business districts of cities and smaller market towns, already hindered by a short supply of developable space.[7] Reaching out to suburbanites where they lived, merchandisers at first built stores along the new highways, in commercial "strips" that consumers could easily reach by car. By the mid-1950s, however, commercial developers, many of whom owned department stores, were constructing a new kind of marketplace, the regional shopping center, aimed at satisfying suburbanites' consumption *and* community needs. Strategically located at highway intersections or along the busiest thoroughfares, the regional

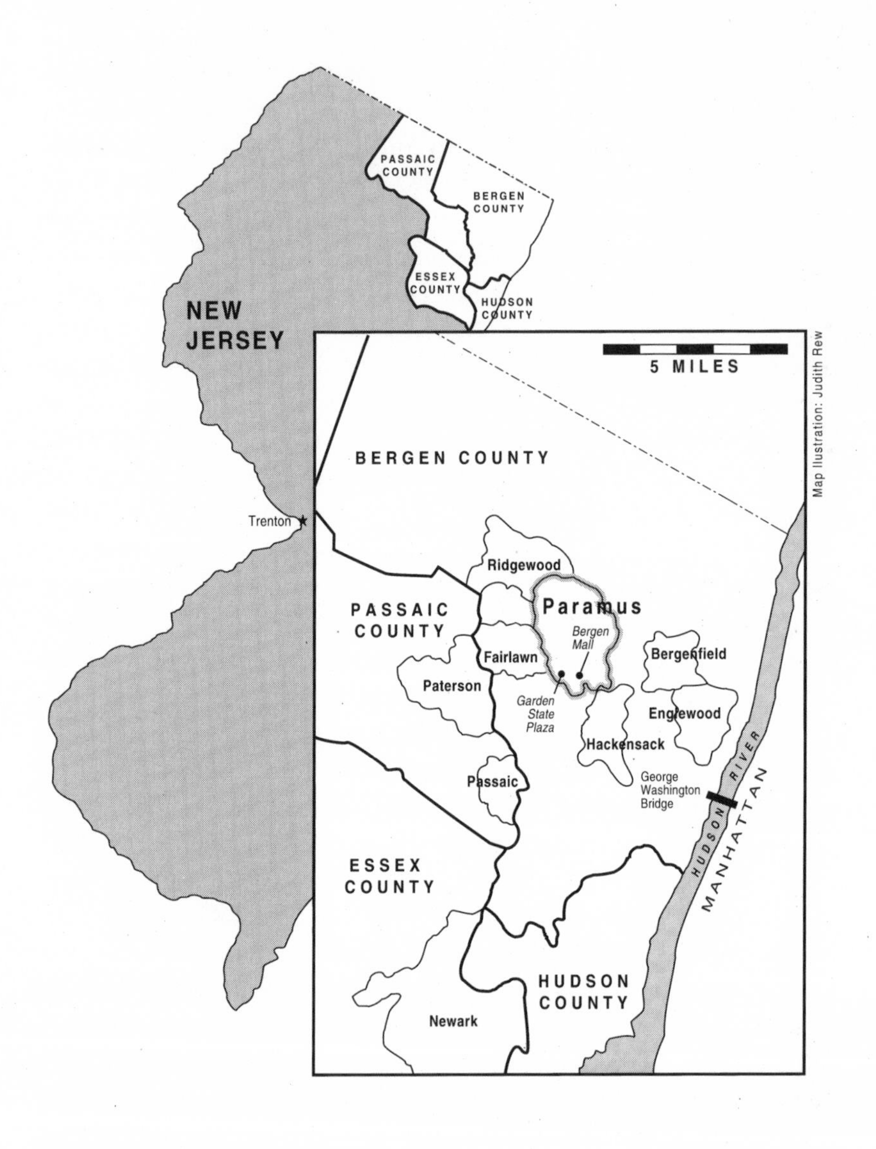

8.1. Location of shopping centers in Paramus, New Jersey, 1957.

shopping center attracted patrons living within half an hour's drive, who could come by car, park in the abundant lots provided, and then proceed on foot (although there was usually some bus service as well). Here was the "new city" of the postwar era, a vision of how community space should be constructed in an economy and society built on mass consumption. Well-designed regional shopping centers would provide the ideal core for a settlement that grew by adding residential nodes off of major roadways rather than concentric rings from downtown, as in cities and earlier suburban communities. After spending several months in the late 1950s visiting these "modern-day downtowns," *Women's Wear Daily* columnist Samuel Feinberg was moved to invoke Lincoln Steffens's proclamation on his return from the Soviet Union in the 1920s: "I have seen the future and it works."[8]

In this chapter I analyze the larger social and political implications of the shift in community marketplace from town center to shopping center. Although I draw on national evidence, I pay special attention to the case of Paramus, New Jersey, a postwar suburb seven miles from the George Washington Bridge that sprouted virtually overnight in the vegetable fields of Bergen County and had become the home of the largest shopping complex in the country by the end of 1957.[9] Within six months R. H. Macy's Garden State Plaza and Allied Stores Corporation's Bergen Mall opened three-quarters of a mile apart at the intersection of Routes 4 and 17 and the soon-to-be-completed Garden State Parkway. Both department store managements had independently recognized the enormous commercial potential of Bergen and Passaic Counties; although the George Washington Bridge connected the area to Manhattan in 1931, the Depression and the war postponed major housing construction until the late 1940s. By 1960 each shopping center had two to three department stores as anchors (distinguishing it from many prewar projects built around a single anchor), surrounded by fifty to seventy smaller stores. Attracting half a million patrons a week, these shopping centers dominated retail trade in the region (figs. 8.1 and 8.2).[10]

The Paramus malls have special significance because of their location adjacent to the wealthiest and busiest central business district in the nation. If these malls could prosper in the shadow of Manhattan, the success of their counterparts elsewhere should come as no surprise.

8.2. This aerial view of the Garden State Plaza in Paramus, New Jersey, shows its orientation to the car. Note its convenient location at the intersection of Routes 4 and 17 and the Garden State Parkway, as well as the extensive space allocated for parking. Courtesy Garden State Plaza Historical Collection.

Moreover, the Paramus case illuminates three major effects of shifting marketplaces on postwar American community life: in commercializing public space they brought to community life the market segmentation that increasingly shaped commerce; in privatizing public space they privileged the rights of private property owners over citizens' traditional rights of free speech in community forums; and in feminizing public space they enhanced women's claim on the suburban landscape but also empowered them more as consumers than as producers.

When planners and shopping-center developers envisioned this new kind of consumption-oriented community center in the 1950s their aim was to perfect the concept of downtown, not to obliterate it, even though their projects directly challenged the viability of existing commercial centers such as Hackensack, the political and commercial seat of Bergen County. It is easy to overlook this visionary dimension and focus only on the obvious commercial motives developers and investors shared. Of course, developers, department-stores owners, and big investors such as insurance companies (who leapt at the promise of a huge return on the vast amounts of capital they controlled) were pursuing the enormous potential for profit in shopping-center development.[11] But they also believed that they were participating in a rationalization of consumption and community no less significant than the way highways were improving transportation or the way tract developments were delivering mass housing.

The ideal was still the creation of centrally located public space that brought together commercial and civic activity. Victor Gruen, one of the most prominent and articulate shopping-center developers, spoke for many others when he argued that shopping centers offered to dispersed suburban populations "crystallization points for suburbia's community life." "By affording opportunities for social life and recreation in a protected pedestrian environment, by incorporating civic and educational facilities, shopping centers can fill an existing void."[12] Not only did Gruen and others promote the construction of community centers in the atomized landscape of suburbia but in appearance their earliest shopping centers idealized—almost romanticized—the physical plan of the traditional downtown shopping street, with stores lining both sides of an open-air pedestrian walkway that was landscaped and equipped with benches (fig. 8.3).[13]

While bringing many of the best qualities of urban life to the suburbs, these new "shopping towns," as Gruen called them, also sought to overcome the "anarchy and ugliness" characteristic of many American cities. A centrally owned and managed Garden State Plaza or Bergen Mall, it was argued, offered an alternative model to the inefficiencies, visual chaos, and provinciality of traditional downtown districts. A centralized administration made possible the perfect mix and "scientific" placement of stores, meeting customers' diverse needs and maximizing store owners' profits. Management kept control visually by standardizing all architectural and

8.3. Garden State Plaza in the early 1960s featured open-air, landscaped walkways, suggesting a pedestrian Main Street. Courtesy Garden State Plaza Historical Collection.

graphic design and politically by requiring all tenants to participate in the tenants' association. Common complaints of downtown shoppers were directly addressed: parking was plentiful, safety was ensured by hired security guards, delivery tunnels and loading courts kept truck traffic away from shoppers, canopied walks and air-conditioned stores made shopping comfortable year-round, piped-in background music replaced the cacophony of the street. The preponderance of chains and franchises over local stores, required by big investors such as insurance companies, brought shoppers the latest national trends in products and merchandising techniques. B. Earl Puckett, Allied Stores' board chair, boasted that Paramus's model shopping centers were making it "one of the first preplanned major cities in America."[14] What made this new market structure so unique and appealing to businessmen like Puckett was that it encouraged social innovation while maximizing profit.

Garden State Plaza and Bergen Mall are good examples of how shopping centers of the 1950s followed Gruen's prescription and became more than miscellaneous collections of stores. As central sites of consumption they offered the full range of businesses and services that one would previously have sought downtown. They not only sold the usual clothing and shoes in their specialty and department stores—Sterns and J. J. Newberry at Bergen Mall, Bamberger's (Macy's New Jersey division), J. C. Penney's, and Gimbels at Garden State Plaza—but also featured stores specifically devoted to furniture, hardware, appliances, groceries, gifts, drugs, books, toys, records, bakery goods, candy, jewelry, garden supplies, hearing aids, tires, even religious objects. Services grew to include restaurants, a post office, a laundromat, a cleaners, a key store, a shoe repair, a bank, a loan company, stock brokerage houses, a barber shop, a travel agency, a real estate office, a "slenderizing salon," and a Catholic chapel. Recreational facilities ranged from a 550-seat movie theater, a bowling alley, and an ice-skating rink to a children's gymnasium and playground.

Both shopping centers made meeting rooms and auditoriums available to community organizations and scheduled a full range of cultural and educational activities to legitimize these sites as civic centers, which also attracted customers. Well-attended programs and exhibitions taught shoppers about such "hot" topics of the 1950s and 1960s as space exploration, color television, modern art, and civics. Evening concerts and

8.4. The in-house publication *Penney News* (Nov.–Dec. 1958) published this photo as part of a feature story on the recent opening of J. C. Penney's Garden State Plaza store. The caption read, "This community club room which can also serve as selling area will be made available free of charge to women's clubs and civic groups." Penney's created this community club room, its first ever, as part of the campaign to make the shopping center the heart of suburban life. P-N.J., Paramus-11. Courtesy J. C. Penney Archives and Historical Museum, Dallas, Texas.

plays, ethnic entertainment, dances and classes for teenagers, campaign appearances by electoral candidates, community outreach for local charities—these were some of the ways that Bergen Mall and Garden State Plaza made themselves indispensable to life in Bergen County. In sum, it was hard to think of consumer items or community events that could not be found at one or the other of these two shopping centers. (In the 1970s a cynical reporter cracked that "the only institution that had not yet invaded" the modern shopping mall was the funeral home.) Furthermore, stores and services were more accessible than those downtown as the centers were open to patrons from 10:00 A.M. to 9:30 P.M., at first four nights a week and by the 1960s six nights a week. In the eyes of a regional planner such as Ernest Erber these postwar shopping centers helped construct a new kind of urbanism appropriate to the automobile age: the "City of Bergen" he named the area in 1960. The *New York Times* agreed, remarking of the Paramus commercial complex, "It lives a night as well as a day existence, glittering like a city when the sun goes down" (fig. 8.4).[15]

When developers and store owners set out to make the shopping center a more perfect downtown they aimed to exclude from this public space unwanted urban groups such as vagrants, prostitutes, racial minorities, and poor people. Market segmentation became the guiding principle of this mix of commercial and civic activity as the shopping center sought, perhaps contradictorily, to legitimize itself as a true community center and to define that community in exclusionary socioeconomic and racial terms. The simple demographics of postwar America helped: when nine of the ten largest cities in the United States lost population between 1950 and 1960 while all metropolitan areas grew, three whites were moving out for every two nonwhites who moved in, laying the groundwork for the racially polarized metropolitan populations of today.[16] In this respect suburbanization must be seen as a new form of racial segregation in the face of a huge wave of African American migration from the South to the North during the 1950s.

Shopping centers did not exclude inadvertently by virtue of their suburban location. Rather, developers deliberately defined their communities through a combination of marketing and policing. Macy's reminded its stockholders in 1955, as it was building its first shopping center, Garden State Plaza, "We are a type of organization that caters primarily to middle-income groups, and our stores reflect this in the merchandise they carry and in their physical surroundings."[17] It was this concern for "physical surroundings" that made the setting of the suburban shopping center appealing to retailers—and ultimately to customers. As Baltimore's Planning Council explained more explicitly than merchants ever would, "Greater numbers of low-income, Negro shoppers in Central Business District stores, coming at the same time as middle and upper income white shoppers are given alternatives in . . . segregated suburban centers, has had unfortunate implications [for downtown shopping].”[18]

Store selection, merchandise, prices, and carefully controlled access to suburban shopping centers supported the class and color line. A survey of consumer expenditures in northern New Jersey in 1960–61 revealed that although 79 percent of all families owned cars, fewer than one-third of those with incomes below $3,000 did, and the low-income population included a higher percentage of nonwhite families than did the population as a whole.[19] Although bus service was available for shoppers without cars, only a tiny proportion arrived that way (in 1966 a daily average of only 600 people came to Garden State Plaza by bus compared with a midweek daily average of 18,000 cars and a holiday peak of 31,000 cars, many carrying more than one passenger), and bus routes were carefully planned to serve nondriving customers, particularly women, from neighboring suburbs, not low-income consumers from cities such as Passaic, Paterson, and Newark.[20] Whereas individual department stores had long targeted particular markets defined by class and race, selling, for example, to “the carriage trade” at the upper end, shopping centers applied market segmentation on the scale of a downtown. In promoting an idealized downtown, shopping centers like Garden State Plaza and Bergen Mall tried to filter out not only the inefficiencies and inconveniences of the city but also the undesirable people who lived there.

If developers and retailers envisioned the regional shopping center as the new American city of postwar suburbia, what actually happened? How successful were shopping centers in attracting patrons and displacing existing urban centers? By investigating the behavior of consumers, on the one hand, and retail businessmen, on the other, we can assess the impact of Bergen Mall and Garden State Plaza on the commercial and community life of Bergen County.

Consumer surveys of the late 1950s and early 1960s, carried out by sociologists and market researchers interested in evaluating the changes wrought by the new regional shopping centers, provide a remarkably good picture of consumer behavior in the era. Before Bergen Mall and Garden State Plaza opened in 1957, Bergen County shoppers satisfied their immediate needs on the main streets of Hackensack and smaller surrounding towns such as Ridgewood, Fair Lawn, Bergenfield, and

Englewood. For more extensive shopping, people went to branches of Sears and Arnold Constable in Hackensack, Meyer Brothers and Quackenbush's department stores in Paterson, Bamberger's, Hahne's, and Kresge's in Newark, and quite often to the big stores in Manhattan. Even before the regional shopping centers opened, the huge influx of new suburban dwellers had raised retail sales in Bergen County from $400 million in 1948 to $700 million in 1954, an increase of 79 percent; by 1958 sales had increased by another 23 percent, to $866 million. Nonetheless, Bergen County residents in 1954 were still spending $650 million outside the county, almost as much as inside.[21]

Samuel and Lois Pratt, professors at Fairleigh Dickinson University, surveyed Bergen County consumers living within a ten-minute drive of the two new shopping centers in 1957, 1958, and 1959 to follow changes in their shopping habits over time. Prior to the opening of the shopping centers seven in ten suburban families surveyed shopped in New York City to some extent. One year after the centers opened the numbers shopping in New York had dropped to six in ten, and a year later fewer than five in ten families shopped there at all. In other words, one-fourth of the entire sample formerly had shopped in New York City but had now entirely stopped. The loss was even more substantial than that; the 15 percent of suburban families who formerly did most of their shopping in New York City—people the Pratts labeled "major shoppers"—showed the sharpest decline" 50 percent by 1958 and 80 percent by 1959. Moreover, those who continued to shop in New York City were spending much less money there; the average annual expenditure in New York by suburban families dropped from $93 to $68 after the regional shopping centers opened. Furthermore, consumers were much less likely to shop in the New York stores that had opened suburban branches; by the end of the first year the number of Bergen County families who had traded in the New York Macy's or Stern's had dropped by half. A similar study of 1,100 shoppers by the New York University School of Retailing confirmed the Pratts' findings: shoppers for women's wear were half as likely to go to New York and a third as likely to go to Hackensack just one year after the shopping centers opened. By the early 1960s a survey of New York area shoppers by a Harvard Business School professor concluded that more than 80 percent of residents of the New Jersey suburbs were most likely to shop close to home for clothing and household items, whereas only 20 percent went most often to Manhattan and 38 percent to New Jersey cities. (Some multiple answers brought the total to more than 100 percent.) Nationwide the trend was the same; retail sales in central business districts declined dramatically between 1958 and 1963, while overall metropolitan sales mushroomed from 10 percent to 20 percent.[22]

The reasons consumers routinely gave for shifting from downtown stores to shopping centers varied, but the overwhelming motivation they articulated was convenience—the ability to drive and park easily, more night hours, improved store layouts, increased self-selection, and simplified credit like the charge plate. The Pratts concluded that shoppers were not so much dissatisfied with New York and Hackensack stores as attracted to the ease and "progressiveness" of shopping-center shopping. People seemed to share the developers' sense that shopping centers were the modern way to consume.[23]

Although overall patronage of stores in surrounding downtowns declined as shopping-center patronage increased, researchers discovered that the story was not so simple; some local stores were benefiting as Bergen County residents became less dependent on New York. Small purchases that shoppers would have made alongside larger ones in New York were now handled closer to home, often in locally owned shops in small downtowns. A large town like Hackensack, however, did not benefit as much as a Ridgewood or an Englewood since it was being displaced as a major shopping site by the shopping centers and its stores were less likely to foster the same kind of loyalty to merchants as shops in small towns. In fact, within a year of the shopping centers' opening, major shoppers used Hackensack a third less; as a consequence, 50 percent of the retail establishments on Main Street reported less business than in the previous year. By 1960, 10 percent of the stores on Hackensack's Main Street had closed because of competition. Bergen County residents were restructuring their consumption patterns by substituting the new shopping centers for New York and for closer, large shopping towns like Hackensack while continuing to shop, mostly for convenience goods and services, in the small town centers near their homes (fig. 8.5).[24]

Although it is hard to evaluate the extent to which people viewed the shopping centers as more than places to shop—as community centers—

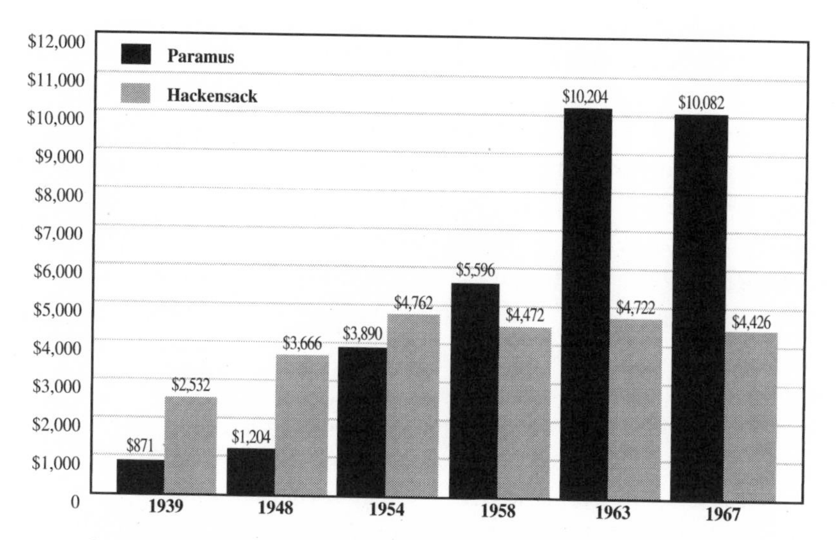

8.5. Per capita annual retail sales, 1939–1967, Paramus and Hackensack, New Jersey. Sources: U.S. Census of Business, Retail Trade Area Statistics, 1939, 1948, 1954, 1958, 1963, 1967; U.S. Bureau of the Census, Census of Population, 1940, 1950, 1960, 1970. Calculated as (total retail sales/consumer price index)/population for nearest year. © Lizabeth Cohen.

anecdotal evidence suggests that they did. Many reporters writing in the late 1950s and 1960s on the way malls were becoming central to the nation's culture made this point, and they routinely introduced their readers to people like Ernest J. Weinhold, a retired designer, who said that he and his wife came to the Cherry Hill Mall in southern New Jersey four days a week. "I love it here—there are things going on that you don't find anywhere else. I don't shop every day but what I do buy I get here."[25] The general manager of Willowbrook Mall, a shopping center not far from Paramus, explained that the Ernest Weinholds of the suburban world made it easy to program activities about forty-five weeks a year. "Whether it's charity fairs, 4-H exhibits, meetings of the Weight-Watchers or the concert by the local barbershop quartet, we find that people respond—and that's what counts."[26] In the new public place of the shopping center consuming and leisure were becoming inseparably intertwined, constructing community experiences around the cultural tastes of white middle-class suburbanites.

The response of businessmen in the existent town centers of northern New Jersey provides further evidence of the crisis confronting local retailers with the arrival of the shopping centers in the late 1950s. As the openings of Bergen Mall and Garden State Plaza neared, and particularly once they were a reality, Main Street retailers in Hackensack, Paterson, and other shopping towns told interviewers that they knew they had to improve their own stores and work cooperatively with other merchants to promote the downtown. In 1957 Hackensack's chamber of commerce launched the first of many campaigns to make shopping in Hackensack more attractive, including the covering of downtown meters at Christmas time so that customers could park free. Paterson, four miles away, formed the Commercial Development Fund for Paterson's Future, which raised $65,000 to promote its downtown through marketing and advertising; at the same time, the Municipal Parking Authority issued $1.8 million in bonds to double the capacity of Paterson's downtown parking lots, raising the total to 800. (By contrast, each shopping center offered parking for more than 8,000 cars.)

Despite the best of intentions to ease parking and traffic, make downtown areas safer, improve customer service, cooperate in promoting downtown shopping, expand merchandise, modernize stores, and strengthen community ties, merchants in Bergen County's town centers had a tough time. Organizing cooperative campaigns of merchants who by identity and practice were independent was extremely difficult. Chambers of commerce and similar entities lacked the coercive, centralized authority of shopping-center managements. Unfortunately, one downtown's promotional campaign was more likely to draw shoppers away from another town in crisis than from the shopping centers. Paterson's gain through the Commercial Development Fund for Paterson's Future, for example, turned out to be neighboring Fair Lawn's loss: retail sales there had dropped to half their 1954 level by the end of 1958. In the specific case of Hackensack, moreover, the Pratts found in a study of

Bergen County retailers that even before the shopping centers opened, the city was in relative decline; although total retail sales grew along with the exploding suburban population, Hackensack's share of the county market decreased from about 20 percent in 1948 to 16 percent in 1954. The shopping centers were only the latest blow to provincial merchants who had not figured out how to prosper in a world undergoing so much social and cultural change.[27]

Recognizing the limits of what they could do alone or through their volunteer merchants' organizations, local Bergen County retailers endorsed two strategies for improving their situation, both of which mobilized the authority and resources of government on their behalf. First, they joined a coalition of other interests, including churches and citizens concerned with traffic congestion, to pass blue laws prohibiting Sunday sales. If the shopping centers were allowed to open on Sunday, small, family-run stores for whom a seven-day week was a great hardship would suffer a handicap. If all stores were required to close, the score would be somewhat even. "It's easy for the big stores to open, but it's different for the independents," explained the owner of a men's clothing store in Hackensack, adding that he and most of his staff of ten worked six days a week. "We are truly a service store, which consists of all full-time people. If you open seven days, you might have to hire part-timers. Our customers want to find a familiar face. They don't want to hear that the person they expect to see is off today."[28] The best defense downtown retailers had against the shopping centers—service—would thus be jeopardized.

Losing no time, Paramus prohibited Sunday sales of virtually all goods except "necessities" (food, drugs, gasoline, newspapers) in 1957, the year the shopping centers opened; violators were subject to a $200 fine or ninety days in jail, or both, per offense, which finally put teeth into a statute long on the state books. The merchants in highway shopping centers protested and sued to have the ordinance revoked, arguing all the way to the New Jersey Supreme Court that their ability to compete with stores in neighboring towns was undermined, but they lost. Meanwhile, agitation continued for an effective statewide restriction of Sunday shopping so as not to penalize particular locales having blue laws, and the New Jersey legislature finally agreed to allow counties to hold referenda on the question. In November 1959 voting took place in fifteen of the state's twenty-one counties; twelve counties, including Bergen, voted a Sunday ban into law. Although highway discount stores appealed, the State Supreme Court eventually upheld the law, as did the U.S. Supreme Court indirectly when it ruled in 1961 on four companion cases concerning the constitutionality of Sunday closing laws in Maryland, Massachusetts, and Pennsylvania. The Supreme Court held that such laws did not violate freedom of religion as protected under the First Amendment or the equal protection guarantees of the Fourteenth Amendment and thereby left it to individual states and localities to regulate Sunday selling as they wished. Due, no doubt, to strong advocacy by influential local businessmen, Bergen County reputedly made the greatest effort to enforce blue laws of any county in New Jersey; a local magistrate even ordered that cigarette vending machines in a Howard Johnson restaurant be unplugged on Sundays. Nationwide during the late 1950s and early 1960s retailers skirmished over Sunday closing laws, not so much defending traditional mores as using the separation of church and state to veil intense struggles over the extent to which discount stores, shopping centers, and chain stores could capture millions of dollars in retail business through restructuring consumer markets.[29]

The second way that downtown business people sought to harness the power of the state in fighting the shopping centers involved the use of federal funds for urban renewal. The 1954 National Housing Act and the 1956 Federal Highway Act made it possible for cities to use urban-renewal grants for rehabilitation of commercial areas; the federal government pledged to pay from two-thirds to three-quarters of the cost of acquiring land and demolishing structures. Paterson proved the most aggressive of the cities in Bergen and Passaic Counties in pursuing this strategy, joining with at least sixteen other communities in the metropolitan New York area. Dissatisfied with the gains from the Commercial Development Fund's promotional and parking efforts, civic leaders founded PLAN (Paterson Looks Ahead Now) in the early 1960s to redevelop 121 acres at the core of the downtown. PLAN implemented a design by Victor Gruen, who had become an early advocate of the revival of downtowns through careful commercial planning, much as he had pioneered the development of regional shopping centers, themselves the source of many cities' economic ills.

Bringing many of the characteristics of Bergen County shopping centers to the Paterson city center, Gruen designed wide, landscaped pedestrian areas, accessible through loop roadways tied in with six parking garages accommodating 4,500 cars. With Uncle Sam committed to footing three-quarters of the $24 million bill, local civic leaders headed by PLAN president Raymond J. Behrman, owner of a downtown luggage and women's accessories store, worked to reverse a drastic decline: by 1962 the number of Paterson shoppers had fallen to half what it had been in 1940, despite all the population growth in the region. Soon Hackensack too was talking about applying for urban-renewal funds. But this injection of federal dollars failed as a remedy. In 1971 shopping centers in the Paterson/Passaic metropolitan area captured 79 percent of all retail trade, well beyond the average of 50 percent for the nation's twenty-one largest metropolitan areas. In 1950 Paterson was a major shopping district, whereas retail in Paramus hardly existed; twenty years later Paterson found itself suffering from long-term economic decline, ignored by recently constructed parkways, turnpikes, and interstates and facing intense competition from shopping centers, whereas Paramus was well on its way to becoming one of the largest retail centers in the world. As the segmentation of consumer markets became the guiding principle in postwar commerce, no amount of revitalization could make a city whose population was becoming increasingly minority and poor attractive to the white middle-class shoppers with money to spend.[30]

While local merchants in Bergen and Passaic Counties struggled, the big New York and Newark stores developed their own strategy for dealing with the competition from the new suburban shopping centers: they opened branch stores. Rather than be eclipsed by the postwar shift in population, they followed it. By the late 1950s, branch stores, once a rarity, had become a national trend among large department stores. Among department stores with annual net sales of $10 million or more the percentage of branch sales skyrocketed from 4 percent of total sales in 1951 to 32 percent by 1959; specialty stores with sales of more than $1 million made a comparable shift, from 6 percent of total sales through branches in 1951 to 33 percent in 1959. By 1959 the very success of a regional shopping center like Bergen Mall or Garden State Plaza depended on the quality of the department-store branches that served as its anchors. In time branch stores evolved from small outlets of Fifth Avenue flagship stores into full-fledged department stores carrying a wide range of merchandise. In the early 1970s, in fact, Bergen Mall's Stern Brothers took the dramatic step of closing its New York City stores, investing everything in its more profitable shopping-center branches. Sterns was not alone; by 1976 branch sales amounted to nearly 78 percent of total department-store business nationwide. The huge postwar investment in suburban stores had significant consequences for consumers, for local retailers, and, as we shall see, for department-store employees as well.[31]

By the 1960s the mass-consumption economy had brought about a major restructuring of consumer markets. As retail dollars moved out of major cities and away from established downtowns within suburban areas, regional shopping centers became the distinctive public space of the postwar landscape. Suburban populations increasingly looked to the mall for a new kind of community life—consumption-oriented, tightly controlled, and aimed at citizen-consumers who preferably were white and middle class. This commercialization of public space during the postwar era had profound effects, perhaps the most important being the struggle to define what kind of political behavior was permissible in the new, privately owned public place.

At first developers sought to legitimize the new shopping centers by arguing for their centrality to both commerce and community, but over time they discovered that those two commitments could be in conflict. The rights of free speech and assembly traditionally safeguarded in the public forums of democratic communities were not always good for business, and they could conflict with the rights of private property owners—the shopping centers—to control entry to their land. Beginning in the 1960s, American courts all the way up to the Supreme Court struggled with the political consequences of having moved public life off the street and into the privately owned shopping center. Shopping centers, in turn, began to reconsider the desirable balance between commerce and community in what had become the major sites where suburbanites congregated.[32]

Once regional shopping centers like the Paramus malls had opened in the 1950s, people began to recognize them as public spaces and to use them to reach out to the community. When the Red Cross held blood drives, when labor unions picketed stores in organizing campaigns, when political candidates campaigned for office, when antiwar and antinuclear activists gathered signatures for petitions, they all viewed the shopping center as the obvious place to reach masses of people. Although shopping centers varied in their responses, from tolerating political activists to monitoring their actions to prohibiting them outright, in general they were wary of any activity that might offend customers. A long, complex series of court tests resulted, culminating in several key Supreme Court decisions that sought to sort out the conflict between two basic rights in a free society: free speech and private property. Not surprisingly, the cases hinged on arguments about the extent to which the shopping center had displaced the traditional "town square" as a legitimate public forum.[33]

The first ruling by the Supreme Court was *Amalgamated Food Employees Union Local 590 v. Logan Valley Plaza, Inc.* (1968), in which Justice Thurgood Marshall, writing for the majority, argued that refusing to let union members picket the Weis Markets in the Logan Valley Plaza in Altoona, Pennsylvania, violated the workers' First Amendment rights since shopping centers had become the "functional equivalent" of a sidewalk in a public business district. Because peaceful picketing and leaflet distribution on "streets, sidewalks, parks, and other similar public places are so historically associated with the exercise of First Amendment rights," he wrote, it should also be protected in the public thoroughfare of a shopping center, even if privately owned. The Logan Valley Plaza decision likened the shopping center to a company town, which had been the subject of an important Supreme Court decision in *Marsh v. Alabama* (1946), upholding the First Amendment rights of a Jehovah's Witness to proselytize in the company town of Chickasaw, Alabama, even though the Gulf Shipbuilding Corporation owned all the property in town. The "Marsh Doctrine" affirmed First Amendment rights over private property rights when an owner opened up his or her property for use by the public.[34] The stance taken in Logan Valley began to give way, however, as the Supreme Court became more conservative under President Richard Nixon's appointees. In *Lloyd Corp. v. Tanner* (1972) Justice Lewis F. Powell Jr. wrote for the majority that allowing antiwar advocates to pass out leaflets at the Lloyd Center in Portland, Oregon, would be an unwarranted infringement of property rights "without significantly enhancing the asserted right of free speech." Antiwar leaflets, he argued, could be effectively distributed elsewhere, without undermining the shopping center's appeal to customers with litter and distraction.[35]

The reigning Supreme Court decision today is *PruneYard Shopping Center v. Robbins* (1980). The Supreme Court upheld a California State Supreme Court ruling that the state constitution granted a group of high school students the right to gather petitions against the U.N. resolution "Zionism Is Racism." The court decided that this action did not violate the San Jose mall owner's rights under the U.S. Constitution. At the same time, however, the court reaffirmed its earlier decisions in *Lloyd v. Tanner* and *Scott Hudgens v. National Labor Relations Board* (1976) that the First Amendment did not guarantee access to shopping malls, and it left it to the states to decide for themselves whether their own constitutions protected such access.

Since *PruneYard*, state appellate courts have been struggling with the issue, and mall owners have won in many more states than they have lost in. In only six states—California, Oregon, Massachusetts, Colorado, Washington, and most recently New Jersey—have state supreme courts protected citizens' right of free speech in privately owned shopping centers. In New Jersey the courts have been involved for some time in adjudicating free speech in shopping centers. In 1983 Bergen Mall was the setting of a suit between its owners and a political candidate who wanted to distribute campaign materials there. When a Paramus Municipal Court judge ruled in favor of the mall, the candidate's attorney successfully appealed on the familiar grounds that "there is no real downtown Paramus. Areas of the mall outside the stores are the town's public sidewalks." He further noted that the mall hosted community events and contained a meeting hall, a post office, and a Roman Catholic chapel. In this case, and in another one the following year over the right of nuclear-freeze advocates to distribute literature at Bergen Mall, free speech was protected on the grounds that the mall was equivalent to a town center.[36]

Such suits should be unnecessary in New Jersey at least for a while, because in a historic decision in December 1994 the New Jersey Supreme Court affirmed that the state constitution guaranteed free speech to opponents of the Persian Gulf War who wanted to distribute leaflets at ten regional malls throughout the state. Writing for the majority, Chief Justice

Robert N. Wilentz confirmed how extensively public space has been transformed in postwar New Jersey:

> The economic lifeblood once found downtown has moved to suburban shopping centers, which have substantially displaced the downtown business districts as the centers of commercial and social activity. . . . Found at these malls are most of the uses and activities citizens engage in outside their homes. . . . This is the new, the improved, the more attractive downtown business district—the new community—and no use is more closely associated with the old downtown than leafletting. Defendants have taken that old downtown away from its former home and moved all of it, except free speech, to the suburbs.

Despite the New Jersey Supreme Court's commitment to free speech, it nonetheless put limits on it, reaffirming the regional mall owners' property rights. Its ruling allowed only the distribution of leaflets—no speeches, bullhorns, pickets, parades, demonstrations, or solicitation of funds. Moreover, the court granted owners broad powers to regulate leaflet distribution by specifying days, hours, and areas in or outside the mall when political activity was permissible. Thus, although shopping centers in New Jersey and five other states have been forced to accommodate some political activity, they have retained authority to regulate it and are even finding ways of preventing legal leafletters from exercising their constitutional rights, such as by requiring them to have million-dollar liability policies, which are often unobtainable or prohibitively expensive. In many other states shopping centers have been able to prohibit political action outright, much as they control the economic and social behavior of shoppers and store owners.[37]

An unintended consequence of the American shift in orientation from public town center to private shopping center, then, has been the narrowing of the ground where constitutionally protected free speech and free assembly can legally take place. As Justice Marshall so prophetically warned in his *Lloyd v. Tanner* dissent in 1972, as he watched the Berger Court reverse many of the liberal decisions of the Warren Court,

> It would not be surprising in the future to see cities rely more and more on private businesses to perform functions once performed by governmental agencies. . . . As governments rely on private enterprise, public property decreases in favor of privately owned property. It becomes harder and harder for citizens to communicate with other citizens. Only the wealthy may find effective communication possible unless we adhere to Marsh v. Alabama and continue to hold that "the more an owner, for his advantage, opens up his property for use by the public in general, the more do his rights become circumscribed by the statutory and constitutional rights of those who use it."[38]

And yet, as Marshall's dissent hinted, and as New Jersey Supreme Court Justice Marie Garibaldi further spelled out, even while advocates for freedom of speech rightfully insist that private property owners respect free speech because their malls have become the new public places, ironically they are endorsing a restructuring of community that could undermine democratic freedom. In the words of Garibaldi's dissent, "Under the majority's theory, private property becomes municipal land and private property owners become the government." The recent increase in the number of self-taxing districts that clean, police, and upgrade neighborhoods, free of municipal oversight or public accountability, suggests some of the worrisome directions in which American society may be headed as once-public spaces and services become privatized.[39]

Along with trends toward commercialization and privatization, the shift from downtown to shopping center entailed a feminization of public space. For at least the last two centuries American women have been the major shoppers in their families. That pattern continued in the postwar period, with marketers estimating that women not only took on anywhere from 80 percent to 92 percent of the shopping but also spent a great deal of their time at it.[40] In a noteworthy departure from earlier times, however, the era of the shopping center saw significant public space—in private hands—being tailored to women's needs and desires as consumers. Although the department store born of the nineteenth century created similarly feminized space, the urban commercial district of which it was a part catered as much to male consumption, leisure, and associational life—through bars, clubs, pool halls, and smoke shops, to say nothing of the male-dominated street resulting from the mix of commercial and corporate culture downtown. The shopping center, in contrast, created the equivalent of a downtown district dedicated primarily to female-orchestrated consumption.[41]

Shopping centers were planned with the female consumer in mind. As women patrons increasingly drove their own cars, they found parking spaces at the shopping center designed wider than usual for the express purpose of making it easier for them—many of whom were new drivers—to park.[42] Women then entered a well-controlled "public" space that made them feel comfortable and safe, with activities planned to appeal especially to women and children. From the color schemes, stroller ramps, baby-sitting services, and special lockers for "ladies' wraps" to the reassuring security guards and special events such as fashion shows, shopping centers were created as female worlds. "I wouldn't know how to design a center for a man," admitted Jack Follet of John Graham, Inc., a firm responsible for many shopping centers. And if New Jersey resident Mrs. Bonnie Porrazzo was any indication, designers like Follet knew what they were doing. Four or five times a week she visited a shopping center three minutes from her suburban home because "it's great for women. What else is there to do?"[43]

Not only did the shopping center pitch itself to women, it sought to empower them as orchestrators of their families' leisure. Marketing surveys revealed that almost half of all women shopped for four or more people, usually members of their own family. With the advent of the suburban malls, they increasingly brought those family members along. Female shoppers in Bergen County surveyed by the Pratts in the first few years after the centers opened revealed that four in ten families were spending more time shopping, three in ten families were making more shopping trips, two in ten women were taking the children more often, and two in ten women were including their husbands more frequently than before the malls were built. A study comparing family shopping in downtown Cincinnati with that in the area's suburban shopping centers concurred, finding that whereas 85 percent of downtown patrons shopped alone, only 43 percent of shopping-center patrons did so; most of the latter were accompanied by family members. Accordingly, evenings and weekends were by far the busiest time in malls, creating peaks and valleys in shopping that had not affected downtown stores nearly as much. In many suburban centers more than half the volume of business was done at night. At Bergen Mall the peak traffic count was at 8:00 P.M., and shopping was very heavy on Saturdays as well. A May Company executive described one of the major problems in branch-store operation: "The biggest day in the suburban store will be ten times the poorest day, instead of five as it usually is downtown."[44]

Shopping centers responded with stores and programming specifically designed to appeal to families in order to encourage them to spend leisure time at the mall in the future. William M. Batten, board chair of the J. C. Penney Company, for example, recalled "the broadening of our lines of merchandise and our services to encompass a fuller spectrum of family activity" as the company began building stores in shopping centers rather than on Main Street in the late 1950s and 1960s; only then the Penney Company start selling appliances, hardware, and sporting goods and offering portrait studios, restaurants, auto service, and Singer sewing instruction. As families strolled and shopped together at the mall they engaged in what increasingly was becoming a form of leisure that was female directed and hence bore witness to a wife's or mother's control (fig. 8.6).[45]

Female authority was also enhanced by shopping centers as they became associated with a huge expansion of consumer credit in the postwar era. In 1950 the ratio of credit to disposable income was 10.4 percent, with $21.5 billion worth of debt outstanding. By 1960 the ratio had grown to 16.1 percent, the debt to $56.1 billion; a decade later these figures had reached 18.5 percent and $127 billion, respectively. The trend was apparent in Bergen County. Bamberger's promoted its Garden State Plaza store as offering "a credit plan to suit every need," a choice between Regular Charge Accounts, Budget Charge Accounts, and Deferred Payment Accounts. Once customers came into the store, an innovative teletype hookup with the Bergen County Credit Bureau enabled charge accounts to be established quickly. Another Garden State Plaza anchor store, J. C. Penney, which had long built its identity around low price, cash-and-carry purchasing, finally recognized in 1957 that credit was expected, even demanded, by consumers and became the last of the large nationwide

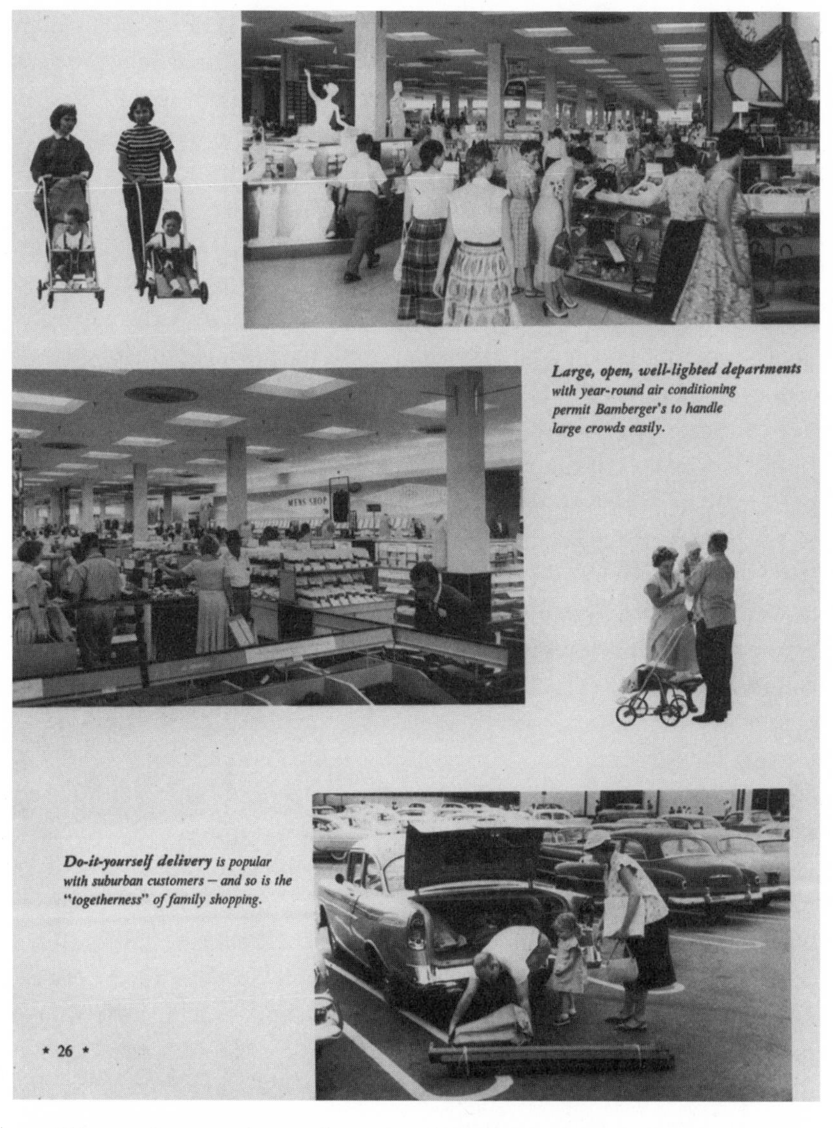

8.6. This page from Macy's annual report to shareholders the year its Garden State Plaza opened conveys the importance of the female-dominated family market to shopping-center merchandising. Note particularly the invocation of "the 'togetherness' of family shopping" in the bottom caption. Reproduced from *Annual Report for 1957*, R. H. Macy & Co., Inc., courtesy Robert F. Wagner Labor Archives, New York University, from its Department Store Workers—Local 1-S Collection, box 2, folder 61.

retailers to introduce a company credit card. By 1962 national credit facilities and systems were operating with the latest electronic data-processing technology, and charging had become the standard way to buy. As credit cards increasingly became the legal tender of shopping-center purchasing they expanded women's control over family finances from spending the domestic allowance assigned from the weekly or monthly paycheck to committing the family's present and future savings. It should also be noted, however, that credit cards at the same time reinforced women's economic dependence on men, since qualifying generally depended on husbands' or fathers' income even when women earned money of their own.[46]

As the example of credit cards illustrates, even as women gained power in the family and in the public realm with the emergence of shopping centers, their horizons were also limited by them. Women's public role was expected to remain defined as that of consumers, and transcending that role was difficult. The most telling case involved the fate of women as workers in shopping centers like Bergen Mall and Garden State Plaza. As the department stores established branches they increasingly turned to suburban housewives to fill positions as retail clerks. The fit seemed perfect: many women were interested in part-time work, and the stores were looking for part-time labor to service the notorious peaks and valleys in suburban shopping. As a Stanford Business School professor advised branch managers in the year the Bergen County shopping centers opened, "Fortunately, most of these suburban stores have in their immediate neighborhood a large number of housewives and other nonemployed women who have been willing to work during these evening and Saturday peak periods. . . . Many of these women apparently work as much because of interest as because of economic necessity, and, as a rule, they have proved to be excellent salespeople." The Paramus malls took heed: by the mid-1960s the part-time employment of women had swelled the malls' combined employee ranks to almost 6,000 people, two-thirds of them part-time and many of them local residents.[47]

But according to New York–area labor unions such as Local 1-S, RWDSU (Retail, Wholesale and Department Store Union), which represented employees at Macy's and Bamberger's, and District 65, RWDSU, representing employees at Gimbels, Sterns, and Bloomingdale's, the department stores had another motive for hiring so many part-timers in their new suburban branches: they were trying to cut labor costs and break the hold of the unions, which had organized their New York stores successfully enough to make retail clerking a decent job. Certainly, retailers gave a lot of attention to keeping labor costs down, judging them to be the greatest obstacle to higher profits. Suburban branch managers sought to limit the number of salespeople needed by depending more on customer self-service and "pipe-racking," putting goods on floor racks rather than behind counters. Some stores, such as Sears Roebuck, Montgomery Ward, and J. C. Penney, expanded their catalog operations.[48] But the basic strategy of the suburban department store was to control wages through hiring more part-timers at minimum wages and benefits.

Organizing the new suburban branches became a life-and-death struggle for the unions beginning in the 1950s. They recognized that not only was the fate of new branch jobs at stake but, as retail dollars left the city for the suburbs, jobs in the downtown stores were threatened as well. The branch store was becoming, in effect, a kind of runaway shop that undermined the job security, wages, benefits, and working conditions of unionized downtown workers. Local 1-S and District 65 tried all kinds of strategies, such as demanding contract coverage of the new branches when renegotiating their existing contracts with downtown stores; getting permission from the National Labor Relations Board to split the bargaining units within particular branch stores (e.g., into selling, nonselling, and restaurant) to facilitate organization; assigning downtown store workers to picket suburban branches during strikes and organizing campaigns; and gaining the right for city-store employees to transfer to branches without losing accumulated seniority and benefits.

But successful labor organization of the suburban branches still proved extremely difficult. Branch-store management at Sterns, Bamberger's, and a Bloomingdale's that opened nearby took an aggressive stand against unionism, harassing and firing employees who showed the least inclination to organize, particularly women. Bill Michelson, executive vice president of District 65, pointed to the mentality of part-time employees as another obstacle to successful organizing: "The part-timer, usually a housewife in a suburban town, is interested in picking up extra money and does not have deep roots in her job." The large turnover among part-time workers—through layoffs as well as voluntary resignation—made organizing them all the harder.[49]

Despite the determined efforts of Local 1-S and District 65 to organize all department-store workers in the Paramus shopping centers, Gimbels was the only store to sign a union contract that covered its Paramus store, and this in exchange for a lesser wage increase and the cancellation of a threatened strike. At all the rest an overwhelmingly female work force worked part-time at minimum wage, with few benefits, no union representation, and limited opportunities for career advancement (fig. 8.7). Work became a way for women to maintain their status as consumers, but it did not significantly empower them as producers who could substantially supplement or be independent of male earnings. At Bamberger's, in fact, the handbook for new employees urged them to use their staff discount to purchase store merchandise (20 percent off for apparel worn on the job, 10 percent on other items) so that they could serve as model consumers for customers. The shopping center, then, contributed to a segmentation not only of consumers but also of workers in a postwar labor market that offered new jobs to women but marked these jobs as less remunerative and more dead-end.[50] Furthermore, as a workplace, much like as a public space, the shopping center constricted the rights available to the people who frequented it. That women came to dominate the ranks of workers and consumers there meant that their political freedom was particularly circumscribed. The shopping center thus posed a contradiction for women in the 1950s and 1960s: it empowered them in their families through creating a new community setting catering to female needs and desires, yet it contained them in the larger society as consumers and part-time workers. In this era before feminist revolt and affirmative action opened other opportunities women's choices were limited not simply through peer pressure and personal priorities, as is often claimed, but also through the larger economic restructuring taking place in the metropolitan marketplace.

Mass consumption in postwar America created a new landscape where public space was more commercialized, more privatized, and more feminized within the regional shopping center than it had been in the traditional downtown center. This is not to romanticize the city and its central business district. Certainly, urban commercial property owners pursued

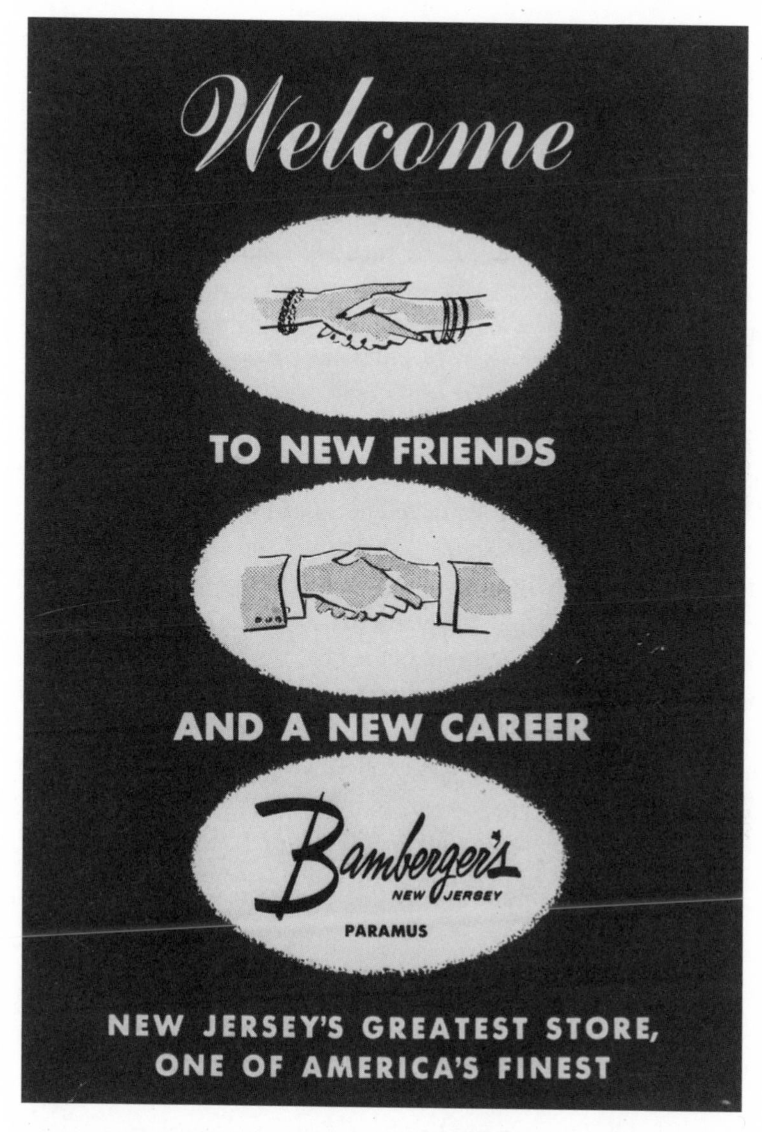

8.7. Bamberger's Department Store prepared an employee handbook, of which this is the cover, for the opening of its Garden State Plaza store in 1957. Hoping to recruit part-time female employees among housewives in neighboring suburban towns, the store offered them "new friends," while male applicants were promised "a new career." Courtesy Robert F. Wagner Labor Archives, New York University, from its Department Store Workers—Local 1-S Collection, box 7, folder 16.

their own economic interests, political activity in public spaces was sometimes limited, and the priorities of women and men did not always peacefully coexist. Nonetheless, the legal distinction between public and private space remained significant; urban loitering and vagrancy laws directed against undesirables in public places have repeatedly been struck down by the courts, whereas privately owned shopping centers have been able to enforce trespassing laws.[51] Overall, an important shift from one kind of social order to another took place between 1950 and 1980, with major consequences for Americans. A free commercial market attached to a relatively free public sphere (for whites) underwent a transformation to a more regulated commercial marketplace (where mall management controlled access, favoring chains over local independents, for example) and a more circumscribed public sphere of limited rights. Economic and social liberalism went hand in hand and declined together.

Not by accident, public space was restructured and segmented by class and race in New Jersey, as in the nation, just as African Americans gained new protections for their right of equal access to public accommodations. Although civil-rights laws had been on the books in New Jersey since the late nineteenth century, comprehensive legislation with mechanisms for enforcement did not pass until the 1940s. With the "Freeman Bill" of 1949, African Americans were finally guaranteed equal access to schools, restaurants, taverns, retail stores, hotels, public transportation, and facilities of commercial leisure such as movie theaters, skating rinks, amusement parks, swimming pools, and beaches, with violators subject to fines and jail terms. Throughout the 1940s and 1950s African American citizens of New Jersey and other northern states vigilantly challenged discrimination by private property owners. Yet larger structural changes in community marketplaces were under way, financed by private commercial interests committed to socioeconomic and racial segmentation. While African Americans and their supporters were prodding courts and legislatures to eliminate legal segregation in public places, real-estate developers, retailers, and consumers were collaborating to shift economic resources to new kinds of segregated spaces.[52]

The landscape of mass consumption created a metropolitan society in which people were no longer brought together in central marketplaces and the parks, streets, and public buildings that surrounded them but, rather, were separated by class, gender, and race in differentiated commercial subcenters. Moreover, all commercial subcenters were not created equal. Over time shopping centers became increasingly class stratified, with some, like Bergen Mall, marketing themselves to the lower middle class, while others, like Garden State Plaza, went upscale to attract upper-middle-class consumers. If tied to international capital, some central business districts—such as New York and San Francisco—have prospered, although they have not been left unscarred by recent retail mergers and leveraged buyouts. Other downtowns, such as Hackensack and Elizabeth, New Jersey, have become "Cheap John Bargain Centers," serving customers too poor and deprived of transportation to shop at malls. Even in larger American cities poor urban populations shop downtown on weekends while the white-collar workers who commute in to offices during the week patronize the suburban malls closer to where they live. Some commercial districts have been taken over by enterprising, often newly arrived ethnic groups, who have breathed new life into what would otherwise have been in decay; nonetheless, they serve a segmented market. Worst off are cities like Newark, once the largest shopping district in the state, which saw every one of its major department stores close between 1964 and 1992 and much of its retail space remain abandoned, leaving residents such as Raymond Mungin to wonder, "I don't have a car to drive out to the malls. What can I do?" Mass consumption was supposed to bring standardization in merchandise and consumption patterns. Instead, diverse social groups are no longer integrated into central consumer marketplaces but rather are consigned to differentiated retail institutions, segmented markets, and new hierarchies.[53]

Finally, the dependence on private spaces for public activity and the more recent privatization of public space gravely threaten the government's constitutional obligations to its citizens. Not only freedom of speech and public assembly in shopping centers are at issue. Just recently Amtrak's Pennsylvania Station in New York City tried to stave off two suits requiring it to respect constitutional rights guaranteed in public places: an effort by artist Michael Lebron to display a political message on the gigantic curved and lighted billboard that he had rented for two months and a suit brought by the Center for Constitutional Rights to force Amtrak to stop ejecting people from the station because they are homeless.[54] When Jürgen Habermas theorized about the rise and fall of a rational public sphere, he recognized the centrality in the eighteenth and nineteenth centuries of accessible urban places—cafés, taverns, coffeehouses, clubs, meeting houses, concert and lecture halls, theaters, and museums—to the emergence and maintenance of a democratic political culture. Over the last half-century transformations in America's economy and metropolitan landscape have expanded the ability of many people to participate in the mass market. But the commercializing, privatizing, and segmenting of physical gathering places that has accompanied mass consumption has made more precarious the shared public sphere upon which our democracy depends.[55]

Notes

I would like to acknowledge the skill and imagination of two research assistants, Deb Steinbach and Susan Spaet. My research was supported by grants from the National Endowment for the Humanities (1993), the American Council of Learned Societies (1994), and New York University (1993–94). I am also grateful to several audiences who shared helpful reactions to versions of this chapter: the international conference "Gender and Modernity in the Era of Rationalization," Columbia University, Sept. 1994; the conference "Significant Locales: Business, Labor, and Industry in the Mid-Atlantic Region," sponsored by the Center for the History of Business, Technology, and Society, Hagley Museum and Library, Oct. 1994; the history department at George Washington University, Feb. 1995; Tricia Rose's American Studies Colloquium, New York University, Feb. 1995; the Historians of Greater Cleveland, May 1995; and my audience at Vassar College, Nov. 1995. Individuals whose readings have especially helped me include Herrick Chapman, Michael Ebner, Ken Jackson, Richard Longstreth, Tricia Rose, David Schuyler, Sylvie Schweitzer, Phil Scranton, and two anonymous readers for the *American Historical Review.*

This chapter first appeared in the *American Historical Review* 101 (Oct. 1996): 1049–81.

1. "The Economy: The Great Shopping Spree," *Time,* 8 Jan. 1965, 58–62 and cover.

2. See Kenneth T. Jackson, *Crabgrass Frontier: The Suburbanization of the United States* (New York: Oxford Univ. Press, 1985); Robert Fishman, *Bourgeois Utopias: The*

Rise and Fall of Suburbia (New York: Basic Books, 1987); Joel Garreau, *Edge City: Life on the New Frontier* (Garden City NY: Doubleday, 1991); William Sharpe and Leonard Wallock, "Bold New City or Built-Up 'Burb? Redefining Contemporary Suburbia," with comments by Robert Bruegmann, Robert Fishman, Margaret Marsh, and June Manning Thomas, *American Quarterly* 46 (Mar. 1994): 1–61; Carol O'Connor, "Sorting Out Suburbia," ibid. 37 (summer 1985): 382–94.

3. The Editors of *Fortune, The Changing American Market* (Garden City NY, 1995), 76.

4. Ann Durkin Keating and Ruth Eckdish Knack, "Shopping in the Planned Community: Evolution of the Park Forest Town Center" (unpublished paper in possession of author); Howard Gillette Jr., "The Evolution of the Planned Shopping Center in Suburb and City," *American Planning Association Journal* 51 (fall 1985): 449–60; Daniel Prosser, "The New Downtowns: Commercial Architecture in Suburban New Jersey, 1920–1970," in *Cities of the Garden State: Essays in the Urban and Suburban History of New Jersey,* by Joel Schwartz and Daniel Prosser (Dubuque IA: Kendall/Hunt, 1977), 113–15; "Park Forest Moves into '52," *House and Home: The Magazine of Building* 1 (Mar. 1952): 115–16; William S. Worley, *J. C. Nichols and the Shaping of Kansas City: Innovation in Planned Residential Communities* (Columbia: Univ. of Missouri Press, 1990); Richard Longstreth, "J. C. Nichols, the Country Club Plaza, and Notions of Modernity," *The Harvard Architecture Review, Vol. 5: Precedent and Invention* (New York: Harvard Architecture Review, 1986), 121–32; William H. Whyte Jr., "The Outgoing Life," *Fortune* 47 (July 1953): 85; Michael Birkner, *A Country Place No More: The Transformation of Bergenfield, New Jersey, 1894–1994* (Rutherford NJ: Fairleigh Dickinson Univ. Press, 1994), 174–77; *Bergen Evening Record,* special Foster Village edition, 10 Aug. 1949.

5. Jackson, *Crabgrass Frontier,* 255–61. On precedents in the pre–World War II period see, by Richard Longstreth, "Silver Spring: Georgia Avenue, Colesville Road, and the Creation of an Alternative 'Downtown' for Metropolitan Washington," in *Streets: Critical Perspectives on Public Space,* ed. Zeynep Celik, Diane Favro, and Richard Ingersoll (Berkeley: Univ. of California Press, 1994), 247–57; "The Neighborhood Shopping Center in Washington, DC, 1930–1941," *Journal of the Society of Architectural Historians* 51 (Mar. 1992): 5–33; and, "The Perils of a Parkless Town," in *The Car and the City: The Automobile, the Built Environment, and Daily Urban Life,* ed. Martin Wachs and Margaret Crawford (Ann Arbor: Univ. of Michigan Press, 1992), 141–53.

6. Editors of *Fortune, Changing American Market,* 78–80, 90. See also "New Need Cited on Store Centers," *New York Times,* 13 Feb. 1955, 7.

7. Richard Longstreth, "The Mixed Blessings of Success: The Hecht Company and Department Store Branch Development after World War II," Occasional Paper No. 14, Jan. 1995, Center for Washington Area Studies, George Washington University.

8. Samuel Feinberg, "Story of Shopping Centers," in *What Makes Shopping Centers Tick* (New York: Women's Wear Daily, 1960), 1, reprinted from *Women's Wear Daily.* For useful background on the development of regional shopping centers see William Severini Kowinski, *The Malling of America: An Inside Look at the Great Consumer Paradise* (New York: William Morrow, 1985); Neil Harris, *Cultural Excursions: Marketing Appetites and Cultural Tastes in Modern America* (Chicago: Univ. of Chicago Press, 1990), 7, 76–77, 278–88; Margaret Crawford, "The World in a Shopping Mall," in *Variations on a Theme Park: The New American City and the End of Public Space,* ed. Michael Sorkin (New York: Hill & Wang, 1992), 3–30; and Gillette, "Evolution of the Planned Shopping Center."

9. On the postwar growth of Paramus and Bergen County see Raymond M. Ralph, *Farmland to Suburbia, 1920–1960,* Bergen County, New Jersey History and Heritage Series, vol. 6 (Hackensack NJ: Bergen County Board of Chosen Freeholders, 1983), 62–71, 76–90; Catherine M. Fogarty, John E. O'Connor, and Charles F. Cummings, *Bergen County: A Pictorial History* (Norfolk VA: Donning, 1985), 182–93; *Beautiful Bergen: The Story of Bergen County, New Jersey* (1962); Patricia M. Ryle, *An Economic Profile of Bergen County, New Jersey* (Trenton: Office of Economic Research, Division of Planning and Research, New Jersey Department of Labor and Industry, Mar. 1980); and League of Women Voters of Bergen County, *Where Can I Live in Bergen County: Factors Affecting Housing Supply* (Closter NJ, 1972).

10. Feinberg, *What Makes Shopping Centers Tick,* 2, 94–102; Ralph, *Farmland to Suburbia,* 70–71, 84–85; Mark A. Stuart, *Our Era, 1960–Present,* Bergen County, New Jersey History and Heritage Series, vol. 7 (Hackensack NJ: Bergen County Board of Chosen Freeholders, 1983), 19–22; Prosser, "New Downtowns," 119–20; Edward T. Thompson, "The Suburb That Macy's Built," *Fortune* 61 (Feb. 1960): 195–200; *Garden State Plaza Merchant's Manual,* 1 May 1957, and certain pages revised in 1959, 1960, 1962, 1963, 1965, 1969, Garden State Plaza Historical Collection, Garden State Plaza Shopping Center.

11. On the financing of shopping centers and the great profits involved see Jerry Jacobs, *The Mall: An Attempted Escape from Everyday Life* (Prospect Heights IL: Waveland Press, 1984), 52.

12. Victor Gruen, "Introverted Architecture," *Progressive Architecture* 38, no. 5 (1957): 204–8; and Victor Gruen and Larry Smith, *Shopping Towns USA: The Planning of Shopping Centers* (New York: Reinhold, 1960), 22–24, both quoted in Gillette, "Evolution of the Planned Shopping Center." For more on Gruen see Kowinski, *Malling of America,* 118–20, 210–14; "Exhibit of Shopping Centers," *New York Times,* 19 Oct. 1954, 42. Paul Goldberger recently profiled shopping-center builder Martin Bucksbaum in "Settling the Suburban Frontier," *New York Times Magazine,* 31 Dec. 1995, 34–35.

13. Robert Bruegmann made the same point about the way the earliest suburban shopping centers resembled downtown shopping streets in a talk to the Urban History Seminar of the Chicago Historical Society, 17 Feb. 1994.

14. B. Earl Puckett, quoted in Feinberg, *What Makes Shopping Centers Tick,* 101. In addition to sources already cited on the control possible in a shopping center compared with that in a downtown see "Shopping Centers Get 'Personality,'" *New York Times,* 29 June 1958, 1.

15. Ernest Erber, "Notes on the 'City of Bergen,'" 14 Sept. 1960, box B, Ernest Erber Papers (hereafter Erber), Newark Public Library (hereafter NPL); "Paramus Booms as a Store Center," *New York Times,* 5 Feb. 1962, 33–34; "The Mall the Merrier, or Is It?" ibid., 21 Nov. 1976, 62. For details on particular stores and activities at Bergen Mall and Garden State Plaza see Feinberg, *What Makes Shopping Centers Tick,* 97–100; Fogarty, O'Connor, and Cummings, *Bergen County,* 189; and Prosser, "New Downtowns," 119. Almost every issue of the *Bergen Evening Record* from 1957 on yields valuable material (in articles and advertisements) on mall stores, services, and activities. The discussion here is based particularly on issues from 8, 13, and 19 Nov. 1957, 8 Jan. 1958, 10 June 1959, and 2 Mar. 1960. See also "Shoppers! Mass Today on Level 1," *New York Times,* 14 June 1994; press release, "The Garden State Plaza Opens Wednesday, May 1st at the Junction of Routes 4 and 17, Paramus," Historical Collection of Garden State Plaza, folder "GSP history"; "It Won't Be Long Now . . . Bamberger's, New Jersey's Greatest Store, Comes to Paramus Soon," promotional leaflet, stamped 22 Aug. 1956, file "Bergen County Shopping Centers," Johnson Free Public Library, Hackensack NJ; and "The Shopping Center," *New York Times,* 1 Feb. 1976, 6–7.

For data on the allocation of shopping-center space in ten regional shopping centers in 1957 see William Applebaum and S. O. Kaylin, *Case Studies in Shopping Center Development and Operation* (New York: International Council of Shopping Centers, 1974), 101. For evidence of the community orientation of shopping centers nationwide see Arthur Herzog, "Shops, Culture, Centers—and More," *New York Times Magazine,* 18 Nov. 1962, 34–35, 109–10, 112–14; and in the *New York Times,* "A Shopping Mall in Suffolk Offering More Than Goods," 22 June 1970, 39; "Supermarkets Hub of Suburbs," 7 Feb. 1971, 58; and "Busy Day in a Busy Mall," 12 Apr. 1972, 55. On the community-relations efforts of branch stores see Clinton L. Oaks, *Managing Suburban Branches of Department Stores* (Stanford: Stanford Univ. Press, 1957), 81–83.

16. George Sternlieb, *The Future of the Downtown Department Store* (Cambridge: Harvard Univ. Press, 1962), 10.

17. R. H. Macy & Company, *Annual Report for 1955* (New York, 1955). The *Times-Advocate,* 14 Mar. 1976, argues that Bamberger's, Macy's store at the Garden State Plaza, was at the forefront of the chain's appeal to the middle- to upper-

income shopper. On market segmentation of shopping centers see also William H. Whyte Jr., *The Organization Man* (New York: Simon & Schuster, 1956), 316–17; Jacobs, *The Mall*, 5, 12; and Albert Bills and Lois Pratt, "Personality Differences among Shopping Centers," *Fairleigh Dickinson University Business Review* 1 (winter 1961): 7-12, which distinguishes between the customers of the Bergen Mall and those of Garden State Plaza in socioeconomic terms. Crawford's "World in a Shopping Mall," in Sorkin, *Variations on a Theme Park*, 8-9, discusses the sophisticated strategies market researchers use to analyze trade areas and pitch stores to different kinds of customers.

18. George Sternlieb, "The Future of Retailing in the Downtown Core," *AIP Journal* 24 (May 1963), as reprinted in Howard A. Schretter, *Downtown Revitalization* (Athens: Univ. of Georgia Press, 1967), 95, and quoted in Jon C. Teaford, *The Rough Road to Renaissance: Urban Revitalization in America, 1940–1985* (Baltimore: Johns Hopkins Univ. Press, 1990), 129.

19. United States Department of Labor, Bureau of Labor Statistics, "Consumer Expenditures and Income, Northern New Jersey, 1960–61," BLS Report No. 237–63, Dec. 1963, Schomburg Center, New York Public Library, Clipping File "Consumer Expenses & Income—NJ."

20. *The Wonder on Routes 4 and 17: Garden State Plaza*, brochure, file "Bergen County Shopping Centers," Johnson Free Public Library; "Notes on Discussion Dealing with Regional (Intermunicipal) Planning Program for Passaic Valley Area (Lower Portion of Passaic Co. and South Bergen)," n.d., box A, folder 3, Erber, NPL; "Memorandum to DAJ and WBS from EE," 22 Nov. 1966, box B, ibid.; National Center for Telephone Research (A Division of Louis Harris and Associates), "A Study of Shoppers' Attitudes toward the Proposed Shopping Mall in the Hudson County Meadowlands Area," conducted for Hartz Mountain Industries, Feb. 1979, Special Collections, Rutgers University, New Brunswick NJ.

21. Stuart, *Our Era*, 20; Lois Pratt, "The Impact of Regional Shopping Centers in Bergen County" (paper delivered at the Conference on the Consequences of Major Economic Innovations in Northern New Jersey, 23 Apr. 1960), in possession of the author.

22. Samuel Pratt and Lois Pratt, "The Impact of Some Regional Shopping Centers," *Journal of Marketing* 25 (Oct. 1960): 44–50; Samuel Pratt, "The Challenge to Retailing" (address to the annual meeting of the Passaic Valley Citizens Planning Association, 24 Apr. 1957), in possession of the author; Pratt, "Impact of Regional Shopping Centers in Bergen County"; Samuel Pratt and James Moran, "How the Regional Shopping Centers May Affect Shopping Habits in Rochelle Park (Preliminary)," in "Business Research Bulletin" (Bureau of Business Research, Fairleigh Dickinson University) 1 (1956, mimeographed); New York University study cited in Thompson, "Suburb That Macy's Built," 196, 200; Regional Plan Associa-

tion, Committee on the Second Regional Plan, "Work Book for Workshops" Conference, Princeton NJ, 25-26 May 1966, box D, pp. V-7 to V-9, Erber, NPL; Stuart U. Rich, *Shopping Behavior of Department Store Customers: A Study of Store Policies and Customer Demand, with Particular Reference to Delivery Service and Telephone Ordering* (Boston: Graduate School of Business Administration at Harvard University, 1963), esp. 133–56, 228; Plan One Research Corporation, New York City, for the Bergen Evening Record Corporation, *The Mighty Market* (Hackensack NJ: Bergen Evening Record, 1971). For national statistics on the decline of retail sales in central business districts while they mushroomed in metropolitan areas between 1958 and 1963 see Teaford, *Rough Road to Renaissance*, 129–31.

23. Pratt and Moran, "How the Regional Shopping Centers May Affect Shopping Habits in Rochelle Park"; Pratt, "Challenge to Retailing," 13–15. For surveys of consumers outside of the New York area see C. T. Jonassen, *Downtown versus Suburban Shopping*, Ohio Marketing Studies, The Ohio State University Special Bulletin Number X-58 (Columbus: Bureau of Business Research, 1953); Sternlieb, *Future of the Downtown Department Store*, 33, 131–33; Rich, *Shopping Behavior of Department Store Customers;* and several important studies described in Pratt, "Challenge to Retailing," 15–19.

24. See all the Pratt studies listed in n. 22, as well as "Hackensack Faces Year of Decision," *Bergen Evening Record*, 10 Jan. 1958, 47.

25. Herzog, "Shops, Culture, Centers—and More," 110, quotation on 114.

26. "The Shopping Centers," *New York Times*, 1 Feb. 1976, 7.

27. The discussion in the two previous paragraphs on the response of local businessmen to shopping-center development is based on "From Now On—Until When?" *Bergen Evening Record*, 6 Dec. 1957, 6; "Bergen Shoppers Shun New York," ibid., 19 Dec. 1957, 1; "Main Street Making Comeback in Duel with Shopping Centers," *New York Times*, 31 May 1962, 1; "Malls Threaten Downtown Suburbia," ibid., 20 Dec. 1972, 92; Samuel Pratt and Lois Pratt, *Suburban Downtown in Transition: A Problem in Business Change in Bergen County, New Jersey* (Rutherford NJ: Institute of Research, Fairleigh Dickinson University, 1958); and Pratt, "Impact of Regional Shopping Centers in Bergen County."

Articles on Hackensack merchants' struggle to compete include, in the *Bergen Evening Record*, "A City with Faith: Hackensack Grows," 29 Oct. 1957, 19; "Alma [Anderson-Linden Merchants Association] Continues to Work to Better Shopping Area," 2 Dec. 1957, 35; "The Way of Alma," 7 Dec. 1957, 28; "Life Line to a City's Future: Hackensack Must Plan, Promote," 10 Jan. 1958, 35; "How to Stimulate Business," 10 Jan. 1958, 42; "What's Ahead for Hackensack Business in '58," 10 Jan. 1958, 48; "So the Fight Is Worth Making," 10 Jan. 1958, 58; "Work Together to Build Business, Chamber Told," 30 Jan. 1958, 3; "Gooding Introduced to City Merchants: First Paid Executive Secretary Will Plan Promotions for Stores," 30 Jan. 1958, 31;

"Bohn Rejects Business Role in Tax Boost," 6 June 1959, 9; and "Main Street Is After Money," *The Record*, 21 Mar. 1968, C1.

On other towns in northern New Jersey see, in the *Bergen Evening Record*, "O'Neil Proposes Park Garage as Boon to Englewood Stores," 18 Oct. 1957, 25; "Village Preparing Spaces for Parking of 90 Cars" (Ridgewood), 6 Nov. 1957, 33; "Chamber to Try Charge-It Plan" (Ramsey), 8 Nov. 1957, 20; "Ridgewood Storekeepers Act to Attract Holiday Shoppers," 12 Nov. 1957, sec. 2, p. 1; "Chamber President Finds Shopping Off" (Bergenfield), 27 Dec. 1957, 9; "Drop of 5–10% Reported in Christmas Business" (Dumont), 31 Dec. 1957, 5; and "Shop at Local Stores, Kiwanis Members Told" (New Milford), 2 Mar. 1960, 9. For an excellent case study of Lancaster, Pennsylvania's struggle with suburban retail competition and central city decline see David Schuyler, "Prologue to Urban Renewal: The Problem of Downtown Lancaster, 1945–1960," *Pennsylvania History* 61 (Jan. 1994): 75–101.

28. "Bergen Wary of Shopping on Sundays," *New York Times*, 7 Sept. 1993.

29. Thompson, "Suburb That Macy's Built," 200; Feinberg, *What Makes Shopping Centers Tick*, 101. The political and legal struggle over establishing blue laws in Paramus and Bergen County can be traced in the *Bergen Evening Record*, beginning in 1957. The battle continued into recent times, with another Bergen County referendum in November 1993, which upheld the Sunday closing ban. Paramus and Bergen County are among a very few places in New Jersey that still have blue laws on the books. On the recent referendum see, in the *New York Times*, "On Sundays, Bergen Shoppers Rest," 7 Dec. 1992; "Bergen Stores Try to Repeal 'Blue Laws,'" 27 Aug. 1993, B1; "Bergen Wary of Shopping on Sundays," 7 Sept. 1993, B1; and "Malls Wrestle with the Blues," 26 Sept. 1993. See also "Both Sides of Aisle Converge over Blue Laws," *The Record*, 13 Oct. 1993, C3; and "Bergen Stores to Stay Closed on Sundays," *Star-Ledger*, 3 Nov. 1993. On the struggle over blue laws nationwide, including in New Jersey, during the late 1950s and 1960s, see Gerald Gunther, *Cases and Materials on Individual Rights in Constitutional Law*, 3rd ed. (Mineola NY: Foundation Press, 1981), 1183–84; and E. B. Weiss, "*Never* on Sunday? A Study on Sunday Retailing" (New York: Doyle Dane Bernbach, 1962, mimeographed), esp. 11, 36–43, 59, 63, 79, 83–84. My thanks to Alexis McCrossen for bringing this last document to my attention.

30. "Supermarkets Hub of Suburbs," *New York Times*, 7 Feb. 1971, 58; "Main Street Making Comeback in Duel with Shopping Centers," ibid., 31 May 1962, 1; Feinberg, *What Makes Shopping Centers Tick*, 100–102; Gillette, "Evolution of the Planned Shopping Center," 454–56; "A New Hackensack Sky Line Looms on Drawing Boards," *Bergen Evening Record*, 26 Dec. 1957, 40; James B. Kenyon, *Industrial Localization and Metropolitan Growth: The Paterson-Passaic District* (Chicago: Univ. of Chicago Press, 1960), 209–10; Paterson Planning Board and Boorman and Dorram, Inc., Consultants, "Traffic and Transportation Survey,

Paterson Master Plan, Report 3, August 1964" (Paterson NJ, 1964); Rutgers University Bureau of Economic Research in contract to the New Jersey Department of Conservation and Economic Development for the Meadowland Regional Development Agency and the State of New Jersey, "Technical Report No. 1H: Patterns of Urban Growth and Decline," Nov. 1966, box A, folder 7, pp. IH-38, 64, Erber, NPL.

31. Sternlieb, *Future of the Downtown Department Store*, 33–36; Rich, *Shopping Center Behavior of Department Store Customers*, 52–54; Editors of *Fortune*, *Changing American Market*, 85–86; John Wallis Johnston, *The Department-Store Buyer: A View from Inside the Parent-Branch Complexes*, Studies in Marketing No. 12 (Austin: Bureau of Business Research, University of Texas at Austin, 1969), 25; Jay Scher, *Financial and Operating Results of Department and Specialty Stores of 1976* (New York: National Retail Merchants Association, 1977), cited in Teaford, *Rough Road to Renaissance*, 208.

32. Shopping centers retreated from promoting themselves as central squares and street corners not only because of the free-speech issue but also to limit the loitering of young people (see, in the *New York Times*, "Supermarkets Hub of Suburbs," 7 Feb. 1971, 58; "Coping with Shopping-Center Crises, Dilemma: How Tough to Get If Young Are Unruly," 7 Mar. 1971, sec. 3, p. 1; and "Shopping Centers Change and Grow," 23 May 1971, sec. 7, p. 1).

33. For a useful summary of the relevant court cases and legal issues involved see Curtis J. Berger, "*PruneYard* Revisited: Political Activity on Private Lands," *New York University Law Review* 66 (June 1991): 633–94; see also "Shopping Centers Change and Grow," *New York Times*, 23 May 1971, sec. 7, p. 1. The corporate shopping center's antagonism to free political expression and social action is discussed in Herbert I. Schiller, *Culture Inc.: The Corporate Takeover of Public Expression* (New York: Oxford Univ. Press, 1989), 98–101.

34. On *Amalgamated v. Logan Valley Plaza* see "Property Rights vs. Free Speech," *New York Times*, 9 July 1972, sec. 7, p. 9; *Amalgamated Food Employees Union Local 590 v. Logan Valley Plaza*, 88 S.Ct. 1601 (1968), *Supreme Court Reporter*, 1601–20; 391 US 308, U.S. Supreme Court Recording Briefs 1967, No. 478, microfiche; and "Free Speech: Peaceful Picketing on Quasi-Public Property," *Minnesota Law Review* 53 (Mar. 1969): 873–82. On *Marsh v. State of Alabama* see 66 S.Ct. 276, *Supreme Court Reporter*, 276–84. Other relevant cases between *Marsh v. Alabama* and *Amalgamated v. Logan Valley Plaza* are *Nahas v. Local 905, Retail Clerks International Assoc.* (1956), *Amalgamated Clothing Workers of America v. Wonderland Shopping Center, Inc.* (1963), and *Schwartz-Torrance Investment Corp. v. Bakery and Confectionary Workers' Union, Local No. 31* (1964); with each case the Warren Court was moving closer to a recognition that the shopping center was becoming a new kind of public forum.

35. "4 Nixon Appointees End Court's School Unanimity, Shopping Centers' Right to Ban Pamphleteering Is Upheld, 5 to 4," *New York Times,* 23 June 1972, 1; "Shopping-Center Industry Hails Court," ibid., 2 July 1972, sec. 3, p. 7; *Lloyd Corporation, Ltd. v. Donald M. Tanner* (1972), 92 S.Ct. 2219 (1972), *Supreme Court Reporter,* 2219–37. The American Civil Liberties Union brief went to great lengths to document the extent to which shopping centers have replaced traditional business districts (see "Brief for Respondents," U.S. Supreme Court Record, microfiche, 20–29; see also "People's Lobby Brief," U.S. Supreme Court Record, microfiche, 5).

The Supreme Court majority wanted to make it clear that in finding in favor of the Lloyd Center it was not reversing the Logan Valley decision, arguing for a distinction based on the fact that antiwar leafletting was "unrelated" to the shopping center, whereas the labor union was picketing an employer. The four dissenting justices, however, were less sure that the distinction was valid and that the Logan Valley decision was not seriously weakened by Lloyd. The important court cases between *Amalgamated v. Logan Valley Plaza* and *Lloyd v. Tanner* include *Blue Ridge Shopping Center v. Schleininger* (1968), *Sutherland v. Southcenter Shopping Center* (1971), and *Diamond v. Bland* (1970, 1974).

36. Berger, "*PruneYard* Revisited"; Kowinski, *Malling of America,* 196–202, 355–59; "Shopping Malls Protest Intrusion by Protesters," *New York Times,* 19 July 1983, B1; "Opening of Malls Fought," ibid., 13 May 1984, sec. 11 (New Jersey), 7; *Michael Robins v. PruneYard Shopping Center* (1979), 592 P. 2nd 341, *Pacific Reporter,* 341–51; *PruneYard Shopping Center v. Michael Robins,* 100 S.Ct. 2035 (1980), *Supreme Court Reporter,* 2035–51; U.S. Supreme Court Record, *PruneYard Shopping Center vs. Robins* (1980), microfiche. The most important Supreme Court case between *Lloyd v. Tanner* and *PruneYard* was *Scott Hudgens v. National Labor Relations Board* (1976), in which the majority decision backed further away from Logan Valley Plaza and refused to see the mall as the functional equivalent of downtown (see *Scott Hudgens v. National Labor Relations Board,* 96 S.Ct. 1029 [1976], *Supreme Court Reporter,* 1029–47).

37. "Court Protects Speech in Malls," *New York Times,* 21 Dec. 1994, A1; "Big Malls Ordered to Allow Leafletting," *Newark Star-Ledger,* 21 Dec. 1994, 1; "Now, Public Rights in Private Domains," *New York Times,* 25 Dec. 1994, E3; "Free Speech in the Mall," ibid., 26 Dec. 1994, 38; Frank Askin, "Shopping for Free Speech at the Malls" (unpublished ms., 1995, in possession of the author).

38. Marshall dissent, *Lloyd v. Tanner,* 92 S.Ct. 2219 (1972), *Supreme Court Reporter,* 2237.

39. See, from the *New York Times,* "Business Districts Grow at Price of Accountability," 20 Nov. 1994, A1; "Now, Public Rights in Private Domains," 25 Dec. 1994, E3; "'Goon Squads' Prey on the Homeless, Advocates Say," 14 Apr. 1995, B1; "City Council Orders Review of 33 Business Improvement Districts," 19 Apr. 1995,

B1; and "When Neighborhoods Are Privatized," 30 Nov. 1995. A 1992 Supreme Court ruling written by Justice Clarence Thomas strengthened the hand of private property owners like the Lechmere store chain in keeping out union organizers. The decision has led malls to ban Salvation Army bell ringers at holiday time in order to protect themselves against union claims to equal access ("A New Grinch Turns Up at the Mall," *New York Times,* 18 Dec. 1995, A12).

40. See Jonassen, *Downtown versus Suburban Shopping,* 15; Alan Voorhees, *Shopping Habits and Travel Patterns* (Washington DC: Urban Land Institute, 1955), 6; Rich, *Shopping Behavior of Department Store Customers,* 61–64; and, on the long history of women as shoppers, Steven Lubar, "Men and Women, Production and Consumption," keynote address to the "His and Hers: Gender and the Consumer" conference, Hagley Museum and Library, Apr. 1994 (a revised version is published as chapter 1 in this volume). The increasingly sophisticated field of market research addressed itself to motivating the female consumer. An excellent example is Janet L. Wolff, *What Makes Women Buy: A Guide to Understanding and Influencing the New Woman of Today* (New York: McGraw-Hill, 1958).

41. My thanks to William Becker and Richard Longstreth, both of George Washington University, for their suggestions on comparing the gendered character of the downtown street with that of the shopping center. See also Gunther Barth, *City People: The Rise of Modern City Culture in Nineteenth-Century America* (New York: Oxford Univ. Press, 1980); Elaine Abelson, *When Ladies Go A-Thieving: Middle-Class Shoplifters in the Victorian Department Store* (New York: Oxford Univ. Press, 1989); and William Leach, "Transformations in a Culture of Consumption: Women and Department Stores, 1890–1925," *Journal of American History* 71 (Sept. 1984): 319–42.

42. On women driving and, specifically, using a car for shopping see Rich, *Shopping Behavior of Department Store Customers,* 84–85, 137–38; Pratt, "Impact of Regional Shopping Centers in Bergen County"; and Voorhees, *Shopping Habits and Travel Patterns,* 17.

43. Herzog, "Shops, Culture, Centers—and More," 35; "Busy Day in Busy Willowbrook Mall," *New York Times,* 22 Apr. 1972, 55, 65; Harris, *Cultural Excursions,* 281.

44. Rich, *Shopping Behavior of Department Store Customers,* 64, 71–74; Pratt, "Impact of Regional Shopping Centers in Bergen County"; Sternlieb, *Future of the Downtown Department Store,* 27–28, 184; Feinberg, *What Makes Shopping Centers Tick,* 97; Oaks, *Managing Suburban Branches of Department Stores,* 72.

45. JCPenney, *An American Legacy: A 90th Anniversary History* (1992, brochure), 22, 25, JCPenney Archives, Dallas; Mary Elizabeth Curry, *Creating an American Institution: The Merchandising Genius of J. C. Penney* (New York: Garland, 1993), 311–13; William M. Batten, *The Penney Idea: Foundation for the Continuing Growth of the J. C. Penney Company* (New York: Newcomen Society in North America, 1967),

17. The opening of the J. C. Penney store in Garden State Plaza in 1958 is featured in the film *The Past Is a Prologue* (1961), one of several fascinating movies made by the company that have been collected on a video, *Penney Premieres,* available through the JCPenney Archives. See also *Penney News* 24 (Nov.–Dec. 1958): 1, 7, on the new Paramus store, JCPenney Archives; and R. H. Macy & Company, *Annual Report for 1957* (New York, 1957), 26.

46. Barry Bluestone, Patricia Hanna, Sarah Kuhn, and Laura Moore, *The Retail Revolution: Market Transformation, Investment, and Labor in the Modern Department Store* (Boston: Auburn House, 1981), 46–47; Rich, *Shopping Behavior of Department Store Customers,* 100–101; "It Won't Be Long Now . . . Bamberger's, New Jersey's Greatest Store, Comes to Paramus Soon"; press release, "The Garden State Plaza Opens Wednesday"; JCPenney, "An American Legacy," 21–22; Curry, *Creating an American Institution,* 305–7.

On the expansion of credit in the postwar period see Marie de Vroet Kobrak, "Consumer Installment Credit and Factors Associated with It" (master's thesis, University of Chicago, 1958); Lewis Mandell, *The Credit Card Industry: A History* (Boston: Twayne, 1990); and Hillel Black, *Buy Now Pay Later* (New York: William Morrow, 1961). For a 1971 study documenting the possession of bank cards and store charge cards in the counties of Bergen, Passaic (New Jersey), and Rockland (New York) and when they were last used see Plan One Research Corporation, *Mighty Market,* 382–85. In 1958 the *Paterson Evening News* cited a recent newspaper poll on family finances showing that the wife had full control of the family purse in 90 percent of all families ("Even as You and I," *Bergen Evening Record,* 10 Jan. 1958, 58; see also "Handling Your Money," *Bergen Evening Record Weekend Magazine,* 25 Jan. 1958, 4).

47. Oaks, *Managing Suburban Branches of Department Stores,* 73; "Paramus Booms as a Store Center," *New York Times,* 5 Feb. 1962, 34; "Sales Personnel Ready to Work," *Bergen Evening Record,* 13 Nov. 1957.

48. Rich, *Shopping Behavior of Department Store Customers,* 20; Sternlieb, *Future of the Downtown Department Store,* 27; R. H. Macy & Company, *Annual Report for 1955,* 29; JCPenney, "An American Legacy," 25; Stuart, *Our Era,* 20.

49. My understanding of labor conditions and organizing in the New York area, and in the Paramus malls specifically, comes from two manuscript collections at the Robert F. Wagner Labor Archives, New York University: the papers of Local 1-S, Department Store Workers' Union (RWDSU), and District 65, now of the UAW, then of the RWDSU. I have based my analysis on the clippings, meeting minutes, and legal files in those collections, which I do not cite individually unless I quote from them. Michelson's statement is quoted from "NLRB Ruling Spurs New York Area Union: Target—50 Stores," box 4, folder 36, District 65; similar statement in "Report to General Council Meeting, Department Store Section, by William

Michelson," 12 Jan. 1965, box 5, folder 4, District 65. On department-store efforts to hire part-time employees see "Part-Timer: New Big Timer," *Women's Wear Daily,* 8 Jan. 1964, box 4, folder 35, Local 1-S; see also see the records of a fascinating case that Local 1-S brought before the NLRB concerning the firing of a young woman employee who had shown interest in the union in box 9, folder 21, Local 1-S.

On industrial relations in department stores nationwide with a case study of the Boston metropolitan area see Bluestone et al., *Retail Revolution,* 70, 80–119, 148–49, which provides an excellent analysis of the restructuring of the labor market in the retail trade. See also Jacobs, *The Mall,* 49.

50. Bamberger's Paramus, *Welcome to New Friends and a New Career,* employee handbook, 1957, box 7, folder 16, pp. 4, 9–12, Local 1-S.

51. See "Amtrak Is Ordered Not to Eject the Homeless from Penn Station," *New York Times,* 22 Feb. 1995, A1.

52. Article on passage of New Jersey Civil Rights Bill, *New York Times,* 24 Mar. 1949; Marion Thompson Wright, "Extending Civil Rights in New Jersey through the Division against Discrimination," *Journal of Negro History* 38 (1953): 96–107; State of New Jersey, Governor's Committee on Civil Liberties, "Memorandum on Behalf of Joint Council for Civil Rights in Support of a Proposed Comprehensive Civil Rights Act for New Jersey," 1948, group 2, series B, container 8, folder entitled "Civil Rights, New Jersey, 1941–48," NAACP Papers, Library of Congress, Washington DC; "Report of Legislative Committee, NJ State Conference of NAACP Branches," 26 Mar. 1949, ibid. Other NAACP files on discrimination document the actual experiences of African Americans in New Jersey during the 1940s and 1950s.

53. "Closing of 'Last' Department Store Stirs Debate on Downtown Trenton," *Star-Ledger,* 5 June 1983; "Urban Areas Crave Return of Big Markets," ibid., 17 July 1984; "Elizabeth Clothier Mourns Demise of Century-Old Customized Service," *Sunday Star-Ledger,* 10 Jan. 1988; "President's Report to the Annual Meeting, Passaic Valley Citizens Planning Association," box A, folder 3, Erber, NPL. On Newark see for the Raymond Mungin quotation "Two Guys Will Be Missed," *Star-Ledger,* 23 Nov. 1981. See also "Last-Minute Bargain Hunters Abound as Chase Closes Up," *Newark News,* 12 Feb. 1967; and "Ohrbach's Will Close Store in Newark, Cites Drop in Sales and Lack of Lease," *New York Times,* 7 Dec. 1973. In the *Star-Ledger* see "S. Klein to Shut Last State Stores Sometime in June," 9 May 1975; "Sears to Shut Newark Store," 13 June 1978; "Hahne's Bids a Farewell to Newark," 18 June 1986; "Macy's to Shut Stores in Newark, Plainfield," 21 May 1992; and "Newmark & Lewis Is Closing 11 Stores," 15 Oct. 1993. And see Greater Newark Chamber of Commerce, "Survey of Jobs and Unemployment," May 1973, and "Metro New Jersey Market Report," 1991, both in "Q" file, NPL.

54. "Amtrak Can Be Sued on Poster, Court Rules," *New York Times,* 22 Feb. 1995,

B4; "Amtrak Is Ordered Not to Eject the Homeless from Penn Station"; "Can Amtrak Be a Censor?" *Washington Post,* 23 Feb. 1995. The Amtrak case was complicated by the ambiguity whether Amtrak is a government entity or a private corporation.

55. Jürgen Habermas, *The Structural Transformation of the Public Sphere: An Inquiry into a Category of Bourgeois Society,* trans. Thomas Burger with Frederick Lawrence (Cambridge: MIT Press, 1989); Geoff Eley, "Nations, Publics, and Political Cultures: Placing Habermas in the Nineteenth Century," in *Culture/Power/History: A Reader in Contemporary Social Theory,* ed. Nicholas B. Dirks, Geoff Eley, and Sherry B. Ortner, (Princeton: Princeton Univ. Press, 1994), 297–335.

About the Contributors

Molly Berger has a doctoral degree from Case Western Reserve University, where she is now a visiting assistant professor of history. She completed her dissertation, "The Modern Hotel in America, 1829–1929," in 1997.

Regina Lee Blaszczyk is an assistant professor of history and American studies at Boston University. Her book *Imagining Consumers: Design, Technology, and Marketing in the Pottery and Glass Industries* is forthcoming from the Johns Hopkins University Press.

Louis Carlat is an editor at the Thomas Edison Papers at Rutgers University. He completed his dissertation, "Sound Values: Radio Broadcasts of Classical Music and American Culture, 1922–1939," at the Johns Hopkins University in 1995.

Lizabeth Cohen is a professor of history at Harvard University. Her book *Making a New Deal: Industrial Workers in Chicago, 1919–1939* (1990) won the Bancroft Prize in 1991. Her chapter is part of a forthcoming book on the impact of mass consumption in America since the 1930s.

Gail Cooper is an associate professor of history at Lehigh University and the author of *Air Conditioning America: American Engineering and the Controlled Environment*, forthcoming from the Johns Hopkins University Press.

Roger Horowitz is associate director of the Center for the History of Business, Technology, and Society at the Hagley Museum and Library and the author of *"Negro and White, Unite and Fight!" A Social History of Industrial Unionism in Meatpacking, 1930–1990* (1997).

Steven Lubar is chair of the Division of the History of Technology at the Smithsonian Institution's National Museum of American History. He has written widely on the history of technology and public history, including *InfoCulture: The Smithsonian Book of Information Age Inventions* (1993).

Arwen Mohun is an associate professor of history at the University of Delaware and the author of *Close to Home: Gender and Technology in the British and American Steam Laundry Industries, 1880–1940*, forthcoming from the Johns Hopkins University Press.

Joy Parr is Farley Professor of History at Simon Fraser University in Vancouver, British Columbia. Her chapter is part of a forthcoming book from the University of Toronto Press, *Domestic Goods: The Material, the Moral, and the Economic in the Postwar Years.*

James C. Williams is a professor of history at De Anza College in Cupertino, California. He has published extensively on the energy industry, most recently *Energy and the Making of Modern California* (1997).

Index

Italicized page numbers refer to illustrations